Flow Injection Atomic Spectroscopy

PRACTICAL SPECTROSCOPY

A SERIES

Edited by Edward G. Brame, Jr.

Crippen Consulting Associates
Wilmington, Delaware

Flow Injection Atomic Spectroscopy

edited by

José Luis Burguera

Department of Chemistry
Faculty of Sciences
University of Los Andes
Mérida, Venezuela

Marcel Dekker, Inc.　　　　New York and Basel

CHEM
cplae

ISBN: 0-8247-8059-0

This book is printed on acid-free paper.

MARCEL DEKKER, INC.
270 Madison Avenue, New York, New York 10016

Current printing (last digit):
10 9 8 7 6 5 4 3 2 1

PRINTED IN THE UNITED STATES OF AMERICA

Foreword

More than a decade has elapsed since the publication of the first paper on flow injection analysis (FIA) (1), and since then the scope of the method has grown at an unprecedented rate. Over 1400 papers on FIA have appeared in scientific journals, with two papers being published each day. The literature on FIA doubles approximately every two years, whereas analytical literature as a whole doubles only after 14 years and general chemical literature after 15. Four monographs have been published so far: *Flow Injection Analysis* (2) in English, Japanese, and Chinese; "Introduction to Flow Injection Analysis, Experiments and Applications" (3) in Japanese; "Fliesinjektionsanalyse" (4) in German and English; *Analysis por Inyeccion en Flujo* (5) in Spanish and English. This monograph and *Flow Injection Analysis 2* (6) are scheduled to appear simultaneously.

This work is, however, unique in two respects: It is the first to deal with a selected field of interest and the first written by an international team of experts and pioneers in the field. Thus, this highly specialized monograph marks the beginning of a new era in the development of such an incredibly versatile technique as FIA. It was wise of the editor to recognize that in the future it will not be possible either to encompass all diverse fields of analytical instrumentation in one monograph or to treat this subject adequately with less than a team of authors. This endeavor will undoubtedly inspire the publication of books dealing with, say, FIA-electrochemistry, FIA-spectrophotometry, FIA-separations, etc. The analytical community comprises many research groups that specialize in selected instrumental techniques, and flow injection analysis will enjoy even broader acceptance if these groups are addressed with their respective fields in mind, in such depth and from so many viewpoints as is the case with this work. No doubt certain topics are treated here more than once. However, from such repetition—inevitable when so many authors, so geographically diverse, write about the same field—the reader will benefit since he or she is given

iii

alternative and perhaps even controversial viewpoints to consider. Different opinions are most valuable in a field undergoing intense development.

I am much honored to be given an opportunity to accompany with this short comment the works of my colleagues, with many of whom I have enjoyed a prior scientific and publishing relationship. I believe that all share my view that the next edition of this monograph will contain many novel FIA techniques not foreseen at this time. No doubt these future discoveries will bring us as much excitement and satisfaction as the discoveries we have shared in the past.

Jaromir Růžička
Technical University of Denmark
Lyngby, Denmark

REFERENCES

1. J. Růžička and E. H. Hansen, *Anal. Chim. Acta, 78,* 145 (1975).

2. J. Růžička and E. H. Hansen, *Flow Injection Analysis,* Wiley, New York, 1981.

3. K. Ueno and K. Kina, "Introduction to Flow Injection Analysis. Experiments and Applications," Kyoto, 1983.

4. J. Moller, "Fliesinjektionsanalyse," *Analytiker Taschenbuch,* Vol. 7, Springer-Verlag, Heidelberg, 1987.

5. M. Valcárcel Cases and M. D. Luque de Castro, *Analysis por Inyeccion en Flujo,* San Pablo Press, Córdoba, Spain (English translation, Ellis Horwood, Chichester, England, 1987).

6. J. Růžička and E. H. Hansen, *Flow Injection Analysis 2,* Wiley, New York, 1988.

Preface

Flow injection analysis is a simple, rapid, and versatile technique that is now firmly established, with widespread application in quantitative chemical analysis. This is apparent from the number of related papers that have appeared in the technical press since 1974.

The three or so previously published books specifically on flow injection analysis treat extensively the fundamentals of the technique and its applications in chemical analysis. During the last few years, many refinements and important related advances have occurred. Therefore, the time has come when it is not only possible but necessary to treat the flow injection literature selectively, and even cautiously.

This book treats in depth and exclusively the fundamentals of flow injection combined with atomic spectroscopy in relation to their significance in chemical analysis. It intends to provide an introduction to the principles and practice of the technique, to serve as a practical handbook on its many experimental aspects, and to bring the reader as up to date as possible with recent developments in the field.

Chapter 1 offers an overview of the subject and indicates the potential of flow injection analysis-atomic spectroscopy (FIA-AS) systems by describing some of the studies and concepts that have evolved in this field. Chapter 2 attempts to rationalize the theoretical background necessary for an adequate understanding of the subject. Chapters 3 to 5 introduce the reader to the fundamental instrumentation, components, and designs of a flow injection-atomic spectroscopy system and its related automation possibilities. In these chapters, special attention is paid to the variability in the fundamental design of flow injection systems from one installation to another and to the different approaches of isolation or the concentration of the elements to be determined. The most important areas of agronomic analysis, clinical analysis, and other kinds of analysis are considered at greater length in Chapters 6 and 7; the final chapter gives an overview of possible future developments in flow injection atomic spectrometry.

This book is written with the analytical chemist and biological scientist in mind. I believe that this volume will also be useful to those who need a rapid orientation in the field and advanced practitioners, helping them to select a particular application in the analysis of various materials.

I wish to acknowledge a pleasant and fruitful relationship, both editorially and professionally, with all my contributors. The patience, assistance, and understanding of the publisher, especially the production editor, Patricia Brecht, are also greatly appreciated. A special acknowledgment is owed to Jaromir Růžička and Miguel Valcárcel, and to my wife, Marcela, for their valuable suggestions.

José Luis Burguera

Contents

Contributors

Henrique Bergamin, Jr. Department of Analytical Chemistry, Center for Nuclear Energy in Agriculture, University of São Paulo, Piracicaba, Brazil

José Luis Burguera Department of Chemistry, Faculty of Sciences, University of Los Andes, Mérida, Venezuela

Marcela Burguera Department of Chemistry, Faculty of Sciences, University of Los Andes, Mérida, Venezuela

Zhaolun Fang Flow Injection Analysis Research Center, Institute of Applied Ecology, Academia Sinica, Shenyang, Liaoning, China

Mercedes Gallego Department of Analytical Chemistry, University of Córdoba, Córdoba, Spain

Francisco José Krug Department of Analytical Chemistry, Center for Nuclear Energy in Agriculture, University of São Paulo, Piracicaba, São Paulo, Brazil

Søren Storgaard Jørgensen Chemistry Department, Royal Veterinary and Agricultural University, Frederiksberg, Copenhagen, Denmark

Gilbert E. Pacey Department of Chemistry, Miami University, Oxford, Ohio

Bernard F. Rocks Biochemistry Department, Royal Sussex County Hospital, Brighton, England

Roy A. Sherwood Biochemistry Department, Royal Sussex County Hospital, Brighton, England

Kent K. Stewart Department of Biochemistry and Nutrition, Virginia Polytechnic Institute and State University, Blacksburg, Virginia

Miguel Valcárcel Department of Analytical Chemistry, University of Córdoba, Córdoba, Spain

Willem E. van der Linden   Department of Chemical Technology, Laboratory for Chemical Analysis, University of Twente, Enschede, The Netherlands

Jacobus F. van Staden   Department of Chemistry, University of Pretoria, Pretoria, South Africa

Elias A. G. Zagatto   Department of Analytical Chemistry, Center for Nuclear Energy in Agriculture, University of São Paulo, Piracicaba, São Paulo, Brazil

1

General Introduction

KENT K. STEWART *Department of Biochemistry and Nutrition, Virginia Polytechnic Institute and State University, Blacksburg, Virginia*

1 INTRODUCTION

Atomic spectroscopy is one of the most sophisticated and elegant methods for direct assay of the monatomic ions. It has the advantages of being specific, sensitive, widely applicable, rapid, and well studied. Somewhere on the order of 70 elements can be determined directly, and many more elements and compounds can be assayed by indirect procedures. Sensitivities are frequently in the parts per million to parts per billion range. The atomic spectroscopic systems of inductively coupled plasma emission spectroscopy and of simultaneous multielement atomic absorption spectroscopy

using a continuous source (SIMACC) (1) permit simultaneous determination of multiple elements over wide ranges of concentrations and
thus expand the versatility of this instrumentation.

Traditional usage of atomic spectroscopy as an assay tool is notable for the amount of manual sample handling usually involved in
the execution of the assay. For example, in the traditional approach
the operator inserts the aspiration tube into the sample container,
waits for the system output signal to reach a steady state, and then
removes the aspiration tube from the sample container. This operation is repeated for each sample. Many assay techniques also require extensive manipulation for sample preparation, concentration,
dilution, and various chemical modifications to alter the chemical
and/or physical characteristics of the analyte, including such activities as precipitation, digestion, extraction, volatilization, and the
formation of soluble complexes. Frequently, when the assay requires a series of complex physical and/or chemical steps, the time
required for individual assays is quite large, and even when batch
approaches are used, the throughputs per day are low. It is not
unusual that the actual measurement of the spectroscopic signal is
the least time-consuming portion of the assay. This extensive reli-
ance on manual manipulation of the sample has numerous undesirable aspects for atomic spectroscopy assays, including a loss of precision, requirements for relatively large sample volumes, a tendency
toward decreased usage of quality control procedures, and the
requirement of a significant amount of operator time for each assay.

The existence of such time-consuming manual operations has led
a number of investigators to examine various approaches when automating the sample handling in atomic spectroscopy. An early approach utilized a combination of segmented continuous flow analysis
with atomic spectroscopy. Although reasonably successful in some
cases, the required air segmentation of the continuous flow systems
can result in air noise in signal from the analyte. The editors and
authors of this book believe that the combination of flow injection
analysis and atomic spectroscopy has greater potential. Our purpose in writing this book is to indicate the potential of flow injection analysis/atomic spectroscopy (FIA-AS) systems by describing
to our readers some of the studies and concepts that have evolved
in this field. It is fitting to start by reviewing briefly the basic
concepts of flow injection analysis.

2   BASIC FLOW INJECTION ANALYSIS

Flow injection analysis (FIA) may be defined as the sequential insertion of discrete sample solutions into an unsegmented continuously
flowing stream with subsequent detection of the analyte (2). Since

its development in the early 1970s, flow injection analysis has grown enormously. Well over 1000 papers and reviews have appeared, over 700 since 1984 (3-11). New papers on all aspects of FIA are appearing with increasing frequency in the analytical chemical literature. Three international conferences have been devoted to FIA, and the Japanese have started a research journal entitled *Journal of Flow Injection Analysis*.

Figure 1 shows a schematic of the basic FIA system for mixing sample and reagents, allowing them to react, and measuring the concentration of the product. Diagrammed is an automated FIA system in which the sample is aspirated from a sample cup in a sampler tray into the sample loop of a sample insertion valve, after which the valve is actuated and the sample is inserted into an unsegmented, continuous stream of sample solvent, which is then mixed with a reagent stream. The resulting mixture then flows through the reaction coil and the detector. Typical FIA recorder tracings are shown in Fig. 2. The analyte concentrations are obtained from either peak heights or peak areas, depending on the individual instrumentation and assay chemistry. With such a FIA system an analyst can perform routine replicate assays of 100 or more samples per hour. In many cases, results for individual samples are obtained within 15 s after the sample is inserted into the system. Precisions of 0.5 to 2.0% RSD have been reported repeatedly for a wide variety of assays. FIA has been used with infrared (IR), visible, and ultraviolet (UV) spectrophotometric, fluorometric, flame emission, atomic absorption, inductively coupled plasma, refractive index, chemiluminescence, thermochemical, and a variety of electrochemical detectors and ion-

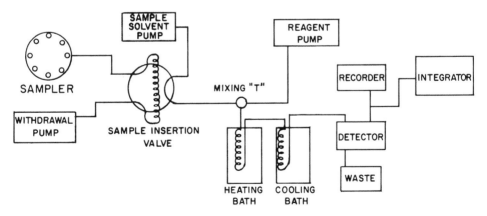

FIGURE 1 Schematic of an automated FIA system. [Reproduced with permission of the American Chemical Society (11).]

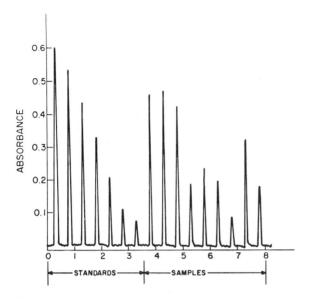

FIGURE 2   Recorder tracing of a FIA determination of zinc.   [Repro-
duced with permission of the American Chemical Society (10).]

sensitive field-effect transistors (11).   There are several variations
on the standard FIA systems shown in Fig. 1, including stopped-
flow FIA and time-based FIA (TBFIA).   Stopped-flow FIA systems
have not been interfaced with atomic spectroscopy systems (for obvious reasons), but time-based (peak width) FIA-AS systems have
been described (12-14).

FIA can be viewed basically as an unsegmented liquid sample
handling system.   Once sequential liquid samples are placed in a
liquid stream, the analyte can be moved, concentrated, diluted,
reacted, purified, and delivered to any detector without interven-
tion of an operator.   Such operations frequently yield assays with
greater accuracy, precision, throughput, and sometimes, better
sensitivity than their manual counterparts.

FIA is frequently compared and/or contrasted with segmented
continuous flow analysis (SCFA) originated by Skeggs and developed by Technicon.   It is this author's opinion that the two systems complement each other as assay techniques.   FIA seems more
suitable for assays with a short reaction time (less than 90 s) and
SCFA seems more suitable for reaction time greater than 1 min.
Each system is capable of providing assays with precisions of 3 to
5% RSD.   Some FIA systems have been reported with precisions
around 0.5% RSD, which seems to be less achievable with SCFA

systems. The lower limit of sample size appears to be about 1 μL for FIA but only about 100 μL for SCFA. On the other hand, SCFA seems to be more suitable for sample sizes of 1 mL and greater. FIA is much more sensitive than SCFA to air bubbles or small particulates. The startup time for FIA seems to be faster than that for SCFA. SCFA techniques are more accepted by industry and by governmental regulatory agencies. FIA seems to be better accepted by academic analytical chemists, but neither technique has been well accepted by academic scientists in the other chemical and biochemical disciplines. Instrument cost, sophistication, and complexity are difficult to compare. The instrumentation for SCFA is quite expensive; that of FIA is less costly. However, the development of SCFA is much further along than that of FIA, particularly in the sophistication and complexity of the control and data handling systems. It is this author's opinion that in the long run, the cost, sophistication, and complexity of SCFA and FIA systems will be about the same. Whereas there are hundreds of existing assays for FIA, there are thousands for SCFA. This is also probably just a sign of the difference in the state of development in the two fields.

In the special case of comparing FIA or SCFA systems with atomic spectrometers as detectors, FIA has several unique advantages. The lack of noise from the air bubbles used in segmented flow and the ability to handle smaller sample volumes are some of the advantages of FIA over SCFA in this particular usage. It would appear that FIA has the potential for much simpler manifolds, the potential of doing more types of in-line chemical modifications and separations of the analyte, and a better potential for miniaturization.

3 FIA-AS SYSTEMS

The usefulness of integrating FIA with atomic spectroscopy was recognized very early in the development of FIA systems. In fact, some of the earliest reports of any systems that would currently be viewed as a FIA system were reported by Sarbeck et al. in 1972 (15) and Chuang et al. in 1973 (16). These workers from Windfordner's group in Florida did injections of microvolume samples into liquid streams which were then aspirated into flame atomic emission or atomic fluorescence spectrophotometers. Reports of FIA combinations with atomic absorbance spectrometers were made in 1979 by three groups (2,17,18), and others reported the interfacing of inductively coupled plasma atomic emission spectrometry with FIA (19). In 1980 there was a brief report of the successful interfacing of a SIMACC system with a FIA system (20). As shown in Fig. 3, after these initial reports, analytical applications of FIA-AS systems have been reported at a steady rate. The author

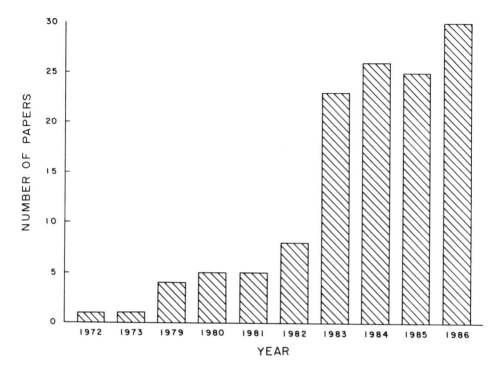

FIGURE 3   Published papers on FIA-AS systems versus year.

currently has in his file over 130 references of studies and reviews on FIA-AS systems, and there are certain to be more by the time this book is published.

4   THEORY

The key component of the theory of FIA systems is the theory of the dispersion of molecules in moving liquid streams.   Analyte samples are customarily inserted into FIA systems in small volumes (usually 10 to 100 μL).   However, as the sample molecules are carried through a FIA system, they are dispersed.   Furthermore, the volume over which the analyte is spread as it goes through the detector increases with increasing reaction time and the length of reaction tubing.   The dispersion can be significant; sample molecules originally contained in 10 to 100 μL are often spread out over several milliliters by the time they reach the detector. The chief causes of this longitudinal dispersion are laminar flow and the discontinuities and various mixing volumes in the FIA system. Control of the dispersion is a critical aspect of FIA, and the theoretical studies on FIA have dealt almost exclusively with the theoretical aspect of the control and manipulation of sample dispersion in FIA systems. Although some aspects of the theory of standard FIA systems are still under development, there is agreement that the flow in these open tube systems is mainly laminar. Furthermore, the longitudinal dispersion of the analyte along the tube is inherent to the laminar flow that occurs in standard FIA systems. The individual peak shapes are influenced by combinations of the laminar flow, molecular diffusion, tube coiling, various system discontinuities, and system mixing chambers (both deliberate and inherent). The details of the theoretical components of standard FIA systems are discussed in detail in Chapter 3 of this volume.

5 SOME BASIC CONCEPTS OF FIA-AS SYSTEMS

Assays involving the insertion of the analyte into moving streams are similar to, but not the same as, reactions in a beaker. It is useful to compare traditional beaker assays with FIA assays to emphasize the basic differences. Although FIA assay systems have been used either to measure the amount of a compound or to measure a catalytic activity, the nature of FIA-AS systems directs the discussion in this chapter to assays dealing with measurements of amounts. These assays have as their critical components control of the amounts (volumes) of sample (analyte), reagents, and carrier (diluent); prevention of individual sample cross-contamination; the mixing of sample, reagent, and carrier; the control of reaction time; and the measurement of amount of product produced.

With classical beaker assays, the analyst will aliquot the amount of sample to be assayed by either weight or volume; in FIA systems, weight is almost never used and the sample is aliquoted into the assay system either by direct volume measurement such as a sample loop in an injection valve or indirectly by the control of a pumping rate and the time of sampling. Similarly in classical beaker assays, the amounts of reagents and diluents can be controlled by weight or volume; in FIA the amounts of the reagent and diluent are usually controlled by pump flow rates, or in the case of merging-zone techniques, by fixed-volume valves. Prevention of individual sample contamination in classical assay is accomplished by the use of clean individual reaction vessels for each analyte sample. In FIA, whereas the samples are usually initially contained in individual containers, the sampling system, reaction system, and detector system

are common to all. Care must be taken that all detectable traces of
each individual sample are removed before the next sample is injected into the system, or the product signal of that sample will be
contaminated by the product signal of the previous sample.

Control of reaction times is crucial to the precision and accuracy
of an assay. Traditionally, the assay time is measured directly.
With FIA, time is not measured directly; instead, the pumping rates,
together with selection of the reaction tubing length and diameter
and the other factors that affect sample dispersion, control the reaction time. Given the potential for precise control of flow rates in
FIA systems, FIA systems frequently have better control of reaction
times than do manual assays, and thus FIA assays can have better
precisions than those of manual assay systems.

Traditionally, in beaker assays the assay solution is physically
stirred and the intensity of the detector signal of solution at physical
equilibrium is used as a measure of the amount of product of an
assay. Measurements are almost never made on unstirred solutions.
Since laminar flow does not provide any mixing, in FIA systems the
sample, reagent, and carrier streams are mixed only by turbulent
flow and by chemical diffusion. Furthermore, in FIA the system is
never completely mixed or at physical equilibrium. Thus in FIA
systems the product concentration is determined by the intensity of
a transient signal produced by a bolus of solute of changing concentration flowing into the detector. The concentration is usually proportional to a peak height or a peak area.

5.1 Common FIA Problems

Given the differences mentioned above, it is not surprising that
some of the common problems of FIA assays differ from those of
manual assays. The common problems of FIA systems are lack of
sensitivity, poor precision, and noise. The problems of lack of
sensitivity are usually due to excessive sample dispersion and incomplete chemical reactions. The excessive sample dispersions are
usually caused by excessive diameters and lengths of reaction tubing; and excessive mixing volumes in sample insertion valves, tubing connectors, and detectors. Such excessive mixing volumes are
usually caused by "eddy currents" and by very low molecular diffusion. The excessive dispersion is best reduced by proper system
design, including minimizing the diameters and length of reaction
tubing and by use of sample insertion valves, tubing connectors,
and detectors with small mixing volumes. The incomplete chemical
reactions are usually due to a combination of the short reaction
times common to FIA (10 to 60 s) and low reaction temperatures.
Increased sensitivity can frequently be achieved by an increase in

reaction temperature. Increases in reaction times are usually less useful, since such increases also lead to increases in dispersion, which tend to decrease sensitivity. With FIA-AS systems, sensitivity can also be improved by changes in the nature of the carrier stream. Significant increases in sensitivity (on the order of tenfold) have been reported when nonaqueous solvents were used to carry the analyte into the atomizer (21,22), presumably due to the higher temperature of the atomizers in such systems.

The problems of poor precision are usually due to variable pumping rates, instability of detectors, long-term "standing waves,"* imprecise sample insertion systems, cyclic reaction bath temperatures, reagent and/or sample degradation, system memory,[†] inappropriate data acquisition systems, or a combination of these factors. Resolution of problems of poor precision is frequently accomplished by a combination of the following: use of pumps with flow rates that are reproducible within ±1%, higher reaction temperatures to ensure that chemical equilibrium is attained, fixed-volume sample insertion systems, stabilization of the reagents, and modification of the chemical composition of the carrier streams or reaction tubing to reduce analyte binding and subsequent system memory.

The choice of the data acquisition system can be crucial in the control of the precision. Two issues need to be addressed: (a) What parameter (peak height or peak area) should be measured? and (b) How shall the data be recorded? There has been a long-standing discussion about the use of peak height or peak area for the determination of analyte concentrations. There is no universal answer; different assays and assay systems can have different precisions for peak height versus peak area measurements. Thus analysts are advised to compare the precision of peak height data versus peak area data for their assay in their assay system. The selection of a data acquisition system can also make a significant

---

*Standing waves with peak-to-peak durations of minutes have been observed in the author's laboratory when multiple steam air pressure pumps were used or when the FIA system had multiple exits to the atmosphere. The imprecision of the systems under study was related directly to the existence or lack of existence of these standing waves. The origin of the standing waves is not well understood.

[†]System memory is frequently caused by weak binding of an analyte to the system components. The analyte is bound and then later released into the reaction system (see Ref. 23). The author has also observed this type of phenomenon with samples containing protein and a Teflon reaction system.

difference in the precision of the results of FIA assays. FIA analyte peaks are frequently of very short time duration. Many recorders and quite a few computer-based data acquisition systems
and data loggers cannot acquire a sufficient number of the data
bytes of the necessary length to yield high-precision assay results.
With such data systems a significant part of the poor precision of
the FIA assay can be due to the data recording. Of crucial importance in the selection of the appropriate data collection system is a
consideration of how the data are recorded for zero-analyte blanks
and samples. The use of traditional chromatographic integrators for
FIA data systems can lead to improper assignment of data to a sample identification number, since most such integrators will not record any value for zero-analyte samples. Thus with those sets of
samples that contain such zero-analyte samples, a mistaken sequence
of results and sample identification can occur.

5.2 Noise

Air bubbles are frequent sources of noise in FIA systems. Air bubbles are usually caused by the outgassing of system solutions caused
by the pressure drop from point of injection to that of detection, or
from the aspiration of air at the fittings, or from injection of air
bubbles with the sample, or from some combination of the above.
Solutions for the latter two sources are straightforward. Solutions
for the outgassing include the degassing of reagents, the replacement of headspace air with a gas such as helium or argon, use of
a debubbling system such as a short segment of thin-walled Teflon
or Gortex tubing, or the use of a back-pressure device. In FIA-
AS systems the interface of the FIA reaction system to the aspiration system of the atomic spectrometer can also be critical to system
sensitivity and precision. The aspiration rate of the carrier stream
into the spectrometer affects both the sensitivity and precision of
the assay (24). There are significant differences of opinions among
the various workers in FIA-AS as to what composes the best aspiration systems, and the reader is advised to be aware of these differences of opinion.
     Thus the basic rules for minimizing problems in FIA-AS systems
are:

1. Maintain precise flow rates.
2. Inject reproducible sample volumes.
3. Minimize the number of fittings, the internal diameter, and
    length of all tubing.
4. Minimize the mixing volumes of the individual components, in-
    cluding the sample insertion valve and detector, thus reducing
    eddy current formation.

5. Reduce all opportunities for air and gas bubble formation.
6. Match the data system specifications to the individual characteristics of the detector and reaction chemistry.
7. Optimize the interface of the FIA system to the aspirator system of the atomic spectrometer.

All of these rules are important. Satisfactory operation of the entire FIA-AS system is dependent on proper integration and operation of each part. One improperly designed or malfunctioning part can seriously degrade the performance of the entire system. FIA-AS system components are discussed in more detail in Chapter 3.

6 TYPES OF FIA-AS ASSAY SYSTEMS

The key steps in atomic spectroscopy assays are sample digestion, chemical modification of the analyte, removal of interfering matrix compounds, delivery of the sample to the spectrometer, data aquisition and manipulation, and data validation procedures. Initial reports of the combination of FIA with atomic spectroscopy dealt with use of the FIA system to deliver a sample to the aspirator of the spectroscopy system. As noted in Chapter 4 of this volume, these FIA-atomic spectrometer systems eliminated the noise problems associated with the air bubbles in segmented sample insertion systems and "partially overcame some of the shortcomings in conventional sample introduction." For example, with such systems an analyst could use sample volumes in the microliter range, could utilize aspiration rates which permitted higher atomization rates in the nebulizer of the spectrometer, and could get high sample throughputs. Use of FIA-AS systems led to a significant depression of matrix effects in the assays. Of particular usefulness was the potential of utilizing samples with salt concentrations, which would clog the burners in conventional systems. Such characteristics make powerful and attractive arguments for the use of FIA-AS systems for sample assay.

6.1 In-Line Chemistries

Soon after the initial reports on the interfacing of FIA to atomic spectrometers came a series of reports on the use of on-line chemical modification and manipulation of the analyte prior to its introduction into the spectrometer. In this series were reported the utilization of in-line physical and chemical systems for sample digestion, concentration, dilution, modification, and cleanup. The utilization of such on-line chemistries in combination with atomic spectroscopy for the assay of various elements is a significant advance in analytical chemistry and is part of the continued transition from manual to automated assays that was pioneered by Stein et al. (25) in the development of automated amino acid analysis, by Skeggs (26) in the development of segmented continuous flow analysis, and by the developers of FIA (2).

6.2 Chemical Modification of the Analyte

Analysts usually modify the analyte and/or its matrix to improve the sensitivity of the assay; to improve the accuracy of the assay by eliminating matrix interferences; to improve the selectivity for the different chemical forms of the analyte of interest; or alternatively, to convert all the different forms of the analyte to one common form. FIA-AS reports on in-line modification have used in-line digestion, precipitation, extraction, ion-exchange chromatography, volatilization, and the formation of various soluble complexes to accomplish these goals.

Sample digestion is frequently the most crucial, time-consuming, and difficult part of most atomic spectrophotometric assays. Frequently, the bulk of the time required for an assay is that for sample digestion. Furthermore, the physical and chemical characteristics of the digestion chambers dictate the use of reasonably large weights of samples,* even when most of the digested sample is not aspirated into the spectrometer. The recently reported on-line microwave digestion of serum samples by Burguera and Burguera (27) indicates that the digestion process can potentially be done on-line for many liquid samples with all of the coincident advantages of small sample volumes, high precision, low limits of detection, high sensitivity, and low cost.[†] The advent of in-line sample digestion is exciting, and this author believes that continued development of such digestion techniques will result in in-line digestion techniques for most of the samples for which trace element assays are needed. Such a family of in-line digestions would be an extremely important addition to the tools available to the analyst doing trace element assays.

Valcárcel and his co-workers have demonstrated that precipitation reactions as separation tools (see Chapter 5) can be used to good effect in FIA-AS systems. Such techniques can be used for direct or indirect assay of various ions, with or without filtration and with or without subsequent dissolution of the precipitate. The

---

*Sample heterogeneity will also dictate the use of large sample weights.

[†]An appropriate approach to the solution of the serious challenges related to the in-line digestion of solid samples still awaits reporting.

reported selectivities for indirect assays of ions such as chloride are impressive. There would appear to be enormous potential for such an approach in principle. It remains to be seen how (and if) the various pragmatic problems associated with precipitates in flowing streams are solved for assays of different species in different matrices. One is intrigued with the potential usefulness given the very large number of traditional assays utilizing precipitations; and equally concerned with the potential difficulties of doing such precipitation assays in FIA systems. The development of this fascinating concept will be followed with interest.

The use of in-line ion-exchange and chelating resins systems for FIA-AS systems have proven to be quite useful. For example, in-line ion-exchange systems have been used for analyte concentration when assaying samples with very low analyte concentrations, as demonstrated by Fang et al. (28,29) in the assay of water samples. In-line ion-exchange systems have also proven to be useful as means of removing interfering ions from the analyte stream. The advantages of this technique are discussed in more detail in Chapter 5. Speciation studies have been done with high-performance liquid chromatography FIA-AS systems as reported by Yoza and Ohishi (30) and by Renoe et al. (31). Liquid-liquid two-phase systems (32) are used to perform many of the same functions as the ion-exchange systems, and in FIA-AS systems, liquid-liquid two-phase systems have proven useful for both analyte concentration and the removal of interfering compounds. The large number of complexing agents and extracting systems makes the liquid-liquid system quite attractive for future method development. The use of ion-exchange and two-phase systems for FIA-AS systems is discussed in more detail in Chapters 4 and 5.

There are several reports of indirect assay of selected species by doing on-line chemistries. In such assays the selective chemistries of the compound being assayed are utilized so that the signal in the atomic spectroscopy detector by a detectable ion is proportional to the original concentration of the original ion, which itself would elicit little or no detector signal. An example is the assay of cyanide by passing the solution through a copper sulfide reactor and then measuring the amount of copper solubilized (33). A different approach is that of enhancement of the signal of a different species by inserting a sample containing the analyte of interest into a reagent containing the detectable species. In this way Martinez-Jiménez et al. (34) developed an indirect assay for aluminum by measuring the increase in an iron signal that resulted from the concurrent aluminum in the solution. Several workers have developed liquid-liquid two-phase systems for indirect assays. These and similar techniques are discussed in detail in Chapters 4 and 5.

In-line generation of volatile metal species such as the hydride generation procedures is well suited to FIA-AS systems (35). Such chemistries yield highly sensitive and selective assays, and the capability of FIA systems for control of the reaction parameters has made this combination quite powerful. See Chapter 4 for further discussion of the use of this powerful and important technique.

As one examines the various chemical and physical manipulations, such as in-line microwave sample digestion, continuous precipitation, liquid-solid extractions, liquid-liquid extractions, gas-liquid systems, ion-exchange chromatography, hydride generation, various gradient techniques, membrane diffusion techniques, and in-line filtrations, one is struck by the potential of the combination of FIA-AS. At a first approximation, it would seem that almost all of the physical and chemical techniques used in manual atomic spectroscopy can be done on-line with FIA-AS systems. The precision of the timing and the control of the other system parameters yield data of good precision. In many ways use of such on-line chemistries has the potential of being even more profound and useful than initial combination of the two systems.

The potential of variations of the FIA-AS systems give it promise for use with many different types of samples. As shown in this book, the technique is suitable for water assays, agricultural samples of all types, clinical samples, and on-line process control. The very small sample volume requirements make the techniques well suited for biotechnology and clinical applications; the high throughputs make the techniques admirably suited for environmental and other situations that generate very large numbers of samples. The system has recently been used for on-line assays of supercritical fluid systems (36).

6.3 New Trends in FIA-AS

The existing features of FIA-AS systems—of small sample volumes, rapid sample throughput, fewer problems with high salt samples, in-line sample concentration, in-line sample treatment, reduced matrix problems, simplification of quality control procedures, and better utilization of expensive instrumentation—by themselves suggest a great future for FIA-AS systems. However, the new trends in FIA-AS research suggest an even brighter future. Along with continued development of the systems described above, there are serious studies presently in progress on the development of instrumentation, such as the miniaturization of FIA-AS systems [FIA systems on a chip (37)], the use of exponential dispersion systems to permit the assay of samples of widely varying concentrations by measurement of dispersed peak widths (time-based FIA systems) (13, 14), the development of self-optimizing FIA systems (38), and the development of high-sensitivity FIA systems. All these developments suggest a strong future for FIA-AS systems. It is the hope of all the authors that this book will stimulate the use of this new approach, and that such use will enable our readers to do more and better assays.

REFERENCES

1. J. M. Harnly, T. C. O'Haver, B. Golden, and W. R. Wolf, Background-corrected simultaneous multi-element atomic absorption spectrometer, *Anal. Chem.*, *51*:2007 (1979).
2. K. K. Stewart, Flow-injection analysis: A review of its early history, *Talanta*, *28*:789 (1981).
3. N. Yoza, FIA bibliography (I), *J. Flow Inject. Anal.*, *1*:32 (1984).
4. N. Yoza, FIA bibliography (II), *J. Flow Inject. Anal.*, *1*:49 (1984).
5. N. Yoza, FIA bibliography (III), *J. Flow Inject. Anal.*, *2*:61 (1985).
6. N. Yoza, FIA bibliography (IV), *J. Flow Inject. Anal.*, *2*:168 (1985).
7. Y. Baba and N. Yoza, FIA bibliography (V), *J. Flow Inject. Anal.*, *3*:49 (1986).
8. T. Imato and M. Watanabe, FIA bibliography (VI), *J. Flow Inject. Anal.*, *4*:40 (1987).
9. T. Imato and H. Ohura, FIA bibliography (VII), *J. Flow Inject. Anal.*, *4*:40 (1987).
10. W. R. Wolf and K. K. Stewart, Automated multiple flow injection analysis for flame atomic absorption spectrometry, *Anal. Chem.*, *51*:1201 (1979).
11. K. K. Stewart, Flow injection analysis, new tool for old assays new approach to analytical measurements, *Anal. Chem.*, *55*:931A (1983).
12. K. K. Stewart and A. G. Rosenfeld, Exponential dilution chambers for scale expansion in flow injection analysis, *Anal. Chem.*, *54*:2368 (1982).
13. J. F. Tyson and J. M. H. Appleton, Flow injection calibration methods for flame atomic absorption spectrometry, *Anal. Proc.*, *20*:260 (1983).
14. J. F. Tyson, Flow injection techniques for extending the working range of atomic absorption spectrometry and U.V.-visible spectrometry, *Anal. Chim. Acta*, *180*:51 (1986).
15. J. R. Sarbeck, P. A. St. John, and J. D. Winefordner, Measurement of microsamples in atomic emission and atomic fluorescence flame spectrometry, *Mikorchim. Acta*, *149*:55 (1972).

16. F. S. Chuang, J. R. Sarbeck, P. A. St. John, and J. D. Winefordner, Flame spectrometric determination of sodium, potassium and calcium in blood serum by measurement of microsamples, *Mikrochim. Acta, 187:*523 (1973).

17. N. Yoza, Y. Aoyagi, and S. Ohashi, Flow injection system for atomic absorption spectrometry, *Anal. Chim. Acta, 111:*163 (1979).

18. E. A. G. Zagatto, F. J. Krug, F. H. Bergamin Fo., S. S. Jørgensen, and B. F. Reis, Merging zones in flow injection analysis. Part 2. Determination of calcium, magnesium and potassium in plant material by continuous flow injection atomic absorption and flame emission spectrometry, *Anal. Chim. Acta, 104:*279 (1979).

19. J. A. C. Broekaert and F. Leis, An injection method for the sequential determination of boron and several metals in wastewater samples by inductively-coupled plasma atomic emission spectrometry, *Anal. Chim. Acta, 109:*73 (1979).

20. W. R. Wolf and J. M. Harnly, Automated simultaneous multielement flame atomic absorption utilizing flow injection analysis: AMFIA-SIMAAC, Abstracts of the Pittsburgh Converence on Analytical Chemistry and Applied Spectroscopy, Atlantic City, N.J., p. 347 (1980).

21. K. Fukamachi and N. Ishibashi, Flow injection atomic absorption spectrometry with organic solvents, *Anal. Chim. Acta, 119:*383 (1980).

22. A. Attiyat and G. D. Christian, Nonaqueous solvents as carrier or sample solvent in flow injection analysis/atomic absorption spectrometry, *Anal. Chem., 56:* 439 (1984).

23. J. M. Harnly and G. R. Beecher, Two-valve injector to minimize nebulizer memory for flow injection atomic absorption spectrometry, *Anal. Chem., 57:*2015 (1985).

24. J. M. Harnly and G. R. Beecher, Signal to noise ratios for flow injection atomic absorption spectrometry, *J. Anal. At. Spectrom., 1:*75 (1986).

25. D. H. Spackman, W. H. Stein, and S. Moore, Automatic recording apparatus for use in the chromatography of amino acids, *Anal. Chem., 30:*1190 (1958).

26. L. T. Skeggs, Automatic method for colorimetric analysis, *Am. J. Clin. Pathol., 28:* 311 (1957).

27. M. Burguera and J. L. Burguera, Flow injection and microwave-oven sample decomposition for determination of copper, zinc and iron in whole blood by atomic absorption spectrometry, *Anal. Chim. Acta, 179:*351 (1986).

28. Z. Fang, S. Xu, and S. Zhang, Determination of trace amounts of nickel by on-line flow injection ion-exchange preconcentration atomic absorption spectrometry, *Fenxi Huaxue, 12:*997 (1984).

29. Z. Fang, J. Růžička, and E. H. Hansen, An efficient flow-injection system with on-line ion-exchange preconcentration for the determination of trace amounts of heavy metals by atomic absorption spectrometry, *Anal. Chim. Acta, 164:*23 (1984).

30. N. Yoza and S. Ohishi, The application of atomic absorption method as a flow detector to gel chromatography, *Anal. Lett.*, *6:*595 (1973).

31. B. W. Renoe, C. E. Shildeler, and J. Savory, Use of a flow-injection sample manipulator as an interface between a "high-performance" liquid chromatograph and an atomic absorption spectrophotometer, *Clin. Chem.*, *27:*1546 (1981).

32. J. C. de Andrade, C. Pasquini, N. Baccan, and J. C. van Loon, Cold vapor atomic absorption determination of mercury by flow injection analysis using a Teflon membrane phase separator coupled to the absorption cell, *Spectrochim. Acta, Part B, 38:*1329 (1983).

33. A. T. Haj-Hussein, G. D. Christian, and J. Růžička, Determination of cyanide by atomic absorption using a flow injection conversion method, *Anal. Chem.*, *58:*38 (1986).

34. P. Martínez-Jiménez, M. Gallego, and M. Valcárcel, Indirect atomic absorption determination of aluminum by flow injection analysis, *Microchem. J.*, *34:*190 (1986).

35. O. Astrom, Flow injection analysis for the determination of bismuth by atomic absorption spectrometry with hydride generation, *Anal. Chem.*, *54:*190 (1982).

36. J. W. Olesik and S. V. Olesik, Supercritical fluid-based sample introduction for inductively coupled plasma atomic spectrometry, *Anal. Chem.*, *59:*796 (1987).

37. J. Růžička, Flow injection analysis from test tube to integrated microconduits, *Anal. Chem.*, *55:*1040 (1983).

38. D. Betteridge, T. J. Sly, and A. P. Wade, Versatile automatic development system for flow injection analysis, *Anal. Chem.*, *58:*2258 (1986).

2

Theoretical Aspects

WILLEM E. VAN DER LINDEN *Department of Chemical Technology, Laboratory for Chemical Analysis, University of Twente, Enschede, The Netherlands*

1 INTRODUCTION

Flow injection analysis (FIA) is a versatile technique that has attracted the attention of scientists from various fields of analytical chemistry, including atomic absorption spectrometry. This is not surprising because FIA is more a method of sample handling than an analytical method as such. The handling starts with injection of the sample plug of an exactly fixed volume in an uninterrupted carrier stream that transports the sample to a detector. On its passage toward the detector, the sample is mixed to some extent with the carrier stream and, depending on the composition of the sample and the nature of the carrier stream, a chemical reaction may proceed. Additional reagents can be introduced at suitable confluence points at any desired place in the system, allowing the performance of fairly complex analytical procedures. To enhance the sensitivity and to diminish the influence of interfering substances, various modules, such as those for extraction and membrane separation, can be incorporated in the manifold as well. Waiting times between the various reagent additions and the final measurement can be adapted by the appropriate choice of length and diameter of the transport lines and the use of properly designed modules, all in relation to the flow rate.

The relatively small sample volumes commonly used in FIA, together with the fact that detection is accomplished in flow-through arrangements, produce signals that exhibit a transient character in which no steady state is reached. For these signals to be analytically meaningful, it is necessary that the entire process of sample transport, mixing, and so on, proceed in a very reproducible way. This statement is tantamount to the one often used in relation to FIA—that the dispersion must be carefully controlled. Dispersion in the context of flowing media is a term that originated in the field of chemical reaction engineering and refers to the observation that not all fluid elements need the same time to pass through the system: There exists a residence time distribution, RTD. The broader the RTD curve, the larger the dispersion.

For the proper design of FIA systems it is desirable to develop guidelines, based on theory, indicating how dispersion can be manipulated. In the discipline of chemical reaction engineering, more than half a century of experience is collected with respect to the factors that control dispersion in transport conduits and reactors of different types. Therefore, it is strongly recommended that the theoretical aspects of FIA on the basis of concepts developed in that discipline (see, e.g., Refs. 1-3) be discussed. Although the dispersion, as observed by the detector, is determined by the contributions of all the component parts of the flow system, attention will first be focused on conduits. In Section 3 the influence of the injection mode, the detector, and other modules is discussed.

1.1 Various Types of Flow

A liquid can flow through a system either in the regular way in which all fluid elements follow streamlines parallel to the main direction of flow, *laminar* flow, or in a random way in which fluid elements continuously change their momentary flow direction but with a net displacement in the bulk flow direction. The latter type of flow will occur at higher flow rates and is denoted as *turbulent*; laminar flow prevails at low flow velocities. The flow rate at which the character will change from laminar to turbulent is determined by the Reynolds number, Re

$$Re = \frac{2R<v>\rho}{\eta} = \frac{2R<v>}{\nu} \tag{1}$$

where

R = conduit radius, m
$<v>$ = mean linear flow velocity, m s^{-1}
ρ = density, kg m^{-3}
η = dynamic viscosity, Pa s
ν = kinematic viscosity, m^2 s^{-1}

For straight conduits with a circular cross section, the flow tends to turbulent for $Re > Re_{crit} \approx 2000$. If the cross section is not circular, the diameter (2R) should be replaced by a hydraulic diameter defined for a conduit as the ratio of four times the cross-sectional area to the wet circumference. The difference in flow character also finds expression in the velocity profiles, as shown in Fig. 1. For laminar flow a parabolic velocity profile is observed, whereas for turbulent flow a more flattened profile is found.

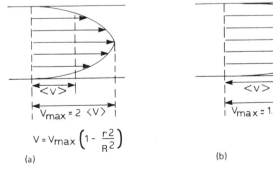

$$V = V_{max}\left(1 - \frac{r^2}{R^2}\right)$$

(a)   (b)

FIGURE 1   Velocity profiles for (a) laminar flow and (b) turbulent flow.

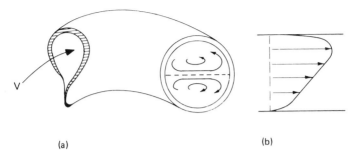

(a)   (b)

FIGURE 2   (a)   Secondary flow in coiled tube; (b) corresponding
velocity profile.

In flow analysis systems the conduits are often coiled.   Such
coiling appears to have a stabilizing effect on laminar flow behavior,
leading to larger values of Re_{crit}.   This effect is caused mainly by
a phenomenon called secondary flow, which is a slow circulation perpendicular to the bulk flow direction.   It originates from the larger
centripetal forces exerted on the faster-moving fluid elements in the
center of the tube, as illustrated in Fig. 2(a).   The corresponding
longitudinal velocity profile is presented as well [Fig. 2(b)].   The
degree of secondary flow is a function of the Dean number, De

$$De = \frac{2R<v>}{\nu}\left(\frac{R}{R_{coil}}\right)^{1/2} = Re\left(\frac{R}{R_{coil}}\right)^{1/2} \qquad (2)$$

where R_{coil} is the radius of the coil (4).
        The velocity profiles depicted in Figs. 1 and 2(b) refer to a
stationary flow.   When a fluid starts to move, some time will pass
before this velocity profile has fully developed.   However, for
aqueous solutions and tube diameters of 0.5 to 1.0 mm, this stabilizing time is for normal flow rates on the order of one-tenth of a
second and hence, in most experiments, neglectable.

1.2   Dimensionless Groups

In the preceding section the Reynolds and Dean numbers were introduced.   These numbers represent dimensionless groups of parameters.   The use of such groups is widespread in chemical reaction engineering because their values are independent of the system
of units used provided that the various parameters within one dimensionless group are consistently expressed in units of the same system.   An additional advantage is the possibility of reducing the
number of variables in a problem, which is of great help in discussing scaling up or down (5).

Dimensionless groups can generally be interpreted as a ratio of two physical phenomena or mechanisms. So it is possible to conceive of the Reynolds number as the ratio of transport of impulse by convection ($\sim \rho vv$) to transport by internal friction ($\sim \eta v/R$).

Other groups of importance in discussing FIA are the Péclet number, the Schmidt number, and "reduced time." The Péclet number (Pe) represents the ratio of mass transport by convection and by diffusion. It is defined as

$$Pe = \frac{\langle v \rangle x}{D} \tag{3}$$

where x is a characteristic length (m) and D the diffusion coefficient ($m^2 \ s^{-1}$). For a tube the characteristic length can be either the length (longitudinal Péclet number, Pe_L) or the diameter (radial Péclet number, Pe_R).

The Schmidt number (Sc) gives the ratio of rates of momentum transport and mass transport

$$Sc = \frac{\nu}{D} \tag{4}$$

It should be noted that

$$Re \cdot Sc = \frac{2R \langle v \rangle}{\nu} \frac{\nu}{D} = \frac{2R \langle v \rangle}{D} = Pe_R \tag{5}$$

Finally, the reduced time (τ) will be used

$$\tau = \frac{t_v D}{R^2} \tag{6}$$

in which t_v is the residence time, defined as $L/\langle v \rangle$, where L is the length of the tube.

1.3 Experimental Domain of FIA

In Table 1 some values are collected for parameters of relevance for FIA experiments. Tube diameters generally range from 0.6 to 1.0 mm and lengths normally vary from 10 to 200 cm. Volumetric flow rates (f_v) are on the order of 0.1 to 3 mL min^{-1}, which yield mean linear flow velocities ranging from about 20 cm s^{-1} ($f_v \sim 3$ mL min^{-1}; R ~ 0.3 mm) to 0.2 cm s^{-1} ($f_v \sim 0.1$ mL min^{-1}; R ~ 0.5 mm). The corresponding Re values will lie in the range 1 to 100, which is well within the region of laminar flow, and consequently the Pe_R (= Re·Sc) values lie between 500 and 50,000. Residence times will, in general, range from 1 to 30 s, which leads to τ values from about 4×10^{-3} to 0.3.

TABLE 1   Data of Relevant Parameters for FIA Experiments

	Solvent 20°C:		
Parameter	H_2O	Methanol	Ethanol
$\eta(N\ s\ m^{-2})$	10^{-3}	1.2×10^{-3}	0.6×10^{-3}
$\nu(m^2\ s^{-1})$	10^{-6}	1.5×10^{-6}	0.75×10^{-6}
$\rho(kg\ m^{-3})$	10^3	0.8×10^3	0.8×10^3
$D(m^2\ s^{-1})$	$\sim2 \times 10^{-9}$	$\sim2 \times 10^{-9}$	$\sim2 \times 10^{-9}$
Sc	~500	~750	~375

Having briefly discussed some general aspects of flowing media, the dispersion of an injected sample plug can now be discussed in more detail.

2   MODELS FOR SAMPLE DISPERSION

Mass transport of a component A in flowing media, taking into account both convective contributions and diffusion, can be described by the following partial differential equation (6):

$$\frac{\partial C_A}{\partial t} + \vec{V} \cdot \nabla C_A = \nabla(\tilde{D} \cdot \nabla C_A) + R_A \tag{7}$$

in which R_A is a source or sink term accounting for the production or disappearance of A, $\vec{V}$ represents the velocity vector, and $\tilde{D}$ is a nonisotropic dispersion coefficient. In this section mass transport without chemical reaction will be discussed, hence $R_A = 0$. In the case of flow through a pipe, bulk flow occurs in the axial direction only, and if we assume radial symmetry, Eq. (7) can be rewritten in cylindrical coordinates

$$\frac{\partial C_A}{\partial t} + V_x(R)\ \frac{\partial C_A}{\partial x} = \frac{\partial}{\partial x}\ D_L(R)\ \frac{\partial C_A}{\partial x} + \frac{1}{R}\ \frac{\partial}{\partial R}\ RD_R(R)\ \frac{\partial C_A}{\partial R} \tag{8}$$

This equation is still too complex to evaluate, and therefore numerical solutions have to be sought or additional restrictions have to be adopted. Usually, the following simplifying assumptions are made; first, the axial and radial dispersion coefficients [$D_L(R)$ and $D_R(R)$, respectively] are assumed to be independent of position:

the *uniform dispersion model*. Second, in addition to the previous simplification, the velocity is taken to be constant and equal to its mean value $\langle v \rangle$. The effect of the velocity profile on dispersion is in this case "lumped" into the dispersion coefficients: the *dispersed plug flow model*. Finally, variations in properties in the radial direction are assumed to be absent. In this *axial-dispersed plug flow model*, mathematical problems are greatly reduced, as can be seen from the corresponding equation

$$\frac{\partial C_A}{\partial t} + \langle v \rangle \frac{\partial C_A}{\partial x} = D'_L \frac{\partial^2 C_A}{\partial x^2} \tag{9}$$

With the usual boundary conditions (i.e., impermeable walls and a double infinite tube), C_A can be calculated as a function of time at $x = L$. For an injection plug of infinitesimal width (perfect pulse injection or delta injection) this yields

$$C_A = \frac{1}{2} C_{A_0} \left(\frac{\langle v \rangle L / D'_L}{\pi \theta} \right)^{1/2} \exp\left[- \frac{(\langle v \rangle L / D'_L)(1 - \theta)^2}{4\theta} \right] \tag{10}$$

where $\theta = t\langle v \rangle / L = t/t_V$. C_{A_0} is defined as the concentration obtained if all the injected A would be homogeneously distributed along the tube, so

$$C_{A_0} = \frac{M_A}{\pi R^2 L} \tag{11}$$

where M_A is the total mass of A injected. Note that $\langle v \rangle L / D'_L$ is a dimensionless group of the Péclet type; this group will be denoted by Pe'_L. In Fig. 3 some C_A/C_{A_0} versus θ curves are drawn. It can be seen that on increasing Pe'_L values the curves become more symmetrical, whereas the maximum approaches $\theta = 1$. For $Pe'_L > 100$ the situation is such that a normal Gauss distribution curve is observed. The standard deviation can be calculated to be

$$\sigma_\theta^2 = \frac{2}{Pe'_L} + \frac{8}{(Pe'_L)^2} \sim \frac{2}{Pe'_L} = \frac{2D'_L}{\langle v \rangle L} \tag{12}$$

or in time units,

$$\sigma_t^2 = \frac{L^2}{\langle v \rangle^2} \sigma_{\theta^2} = \frac{2D'_L L}{\langle v \rangle^3} \tag{13}$$

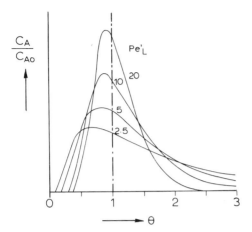

FIGURE 3 Response curves on perfect pulse injection for various
values of the Péclet number based on axially dispersed plug flow
model.

For smaller values of Pe'_L the maximum will be located at $\theta < 1$,
whereas the asymmetry (i.e., a tailing at the high θ end) increases.

Although the theoretically derived curves perfectly match with
the experimental ones, provided that the appropriate D'_L value is
used, there is a serious drawback: D'_L bears no direct relation to
geometrical or physical parameters and can only be determined in
an experimental way. Fortunately, Taylor (7) was able to show
that under certain conditions it is possible to correlate D'_L with the
diffusion coefficient, the tube radius, and the mean linear flow velocity. This "Taylor flow" is discussed in a qualitative way in the
next section. Furthermore, it was found that response curves similar to those shown in Fig. 3 are obtained if a sample plug is injected in a fluid flowing through a series of well-stirred tanks.
Therefore, some attention will also be paid to the tanks-in-series
model.

2.1 Taylor Flow

If convective transport would be the only operating mechanism by
which a sample plug is transported through a conduit, the various
fluid elements will cover a distance depending on the flow velocity
on the respective streamlines. The sample concentration profiles
developed inside the tube are shown for two consecutive times in
Fig. 4(a). The response curve observed at the end of the conduit
will exhibit a severe tailing [Fig. 4(b)].

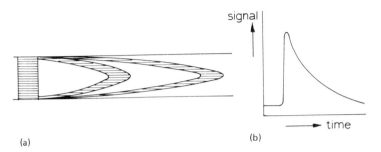

(a)   (b)

FIGURE 4   (a) Development of concentration profile for laminar flow without diffusion; (b) corresponding response curve.

However, at increasing residence times (i.e., at lowering the flow rate), diffusion will start to play a more prominent role. More specifically the radial diffusion component is of importance because it causes the molecules to move to fluid elements at different streamlines. When the total residence time is long enough, each individual molecule will spend approximately the same fraction of time on each streamline and the average velocity of almost all molecules will become equal apart from small statistical variations. The resulting RTD is more or less Gaussian shaped, with a maximum at $t_V = L/\langle v \rangle$. In fact, what is described here in a qualitative way forms the justification of the use of the axial-dispersed plug flow in a pipe. Elaborating this concept of radial smoothing of concentration differences by diffusion, Taylor (7) was able to derive that this type of dispersion will occur if

$$t_V \gg \frac{R^2}{(3.8)^2 D} \quad \text{or} \quad Re \cdot Sc \ll 30 \frac{L}{2R} \tag{14}$$

Expressed in reduced time units, this criterion reads $\tau \gg 0.07$. Furthermore, he could find an expression for D'_L

$$D'_L = \frac{R^2 \langle v \rangle^2}{48D} \tag{15}$$

Substitution in Eq. (13) yields

$$\sigma_t^2 = \frac{t_V R^2}{24D} \tag{16}$$

More detailed studies by other authors have demonstrated that the criterion can be specified to $\tau \geq 0.5$ (8-10).

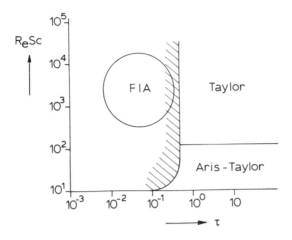

FIGURE 5 Plot showing the various regions where theoretical solutions have been found together with FIA region.

An extension of Taylor's approach was presented by Aris (11), who found that lowering the mean linear flow velocity (i.e., decreasing $Pe_R = Re \cdot Sc$) eventually leads to a situation in which the axial diffusion no longer can be neglected. This leads to the Aris-Taylor equation

$$D_L' = D + \frac{R^2 \langle v \rangle^2}{48D} \tag{17}$$

The domains where Taylor and Aris-Taylor equations are applicable are indicated in Fig. 5. Also indicated is the area in which most of the practical FIA experiments are located (see Section 1.3). Obviously, most FIA applications are outside the Taylor region.

Extensive numerical methods for solutions of the diffusion-convection equation have been given by Ananthakrishnan (10) and Bate et al. (12,13), in particular for the shaded area next to the Taylor domain. Although their results are not very easy to interpret for the development of guidelines, Vanderslice et al. (14) were able to derive the following set of correlations for the time elapsed between injection and the initial appearance of a peak at the detector (t_A) and the peak width (Δt_B; baseline to baseline):

$$t_A = \frac{109(2R)^2 D^{0.025}}{f} \left(\frac{L}{f_v} \right)^{1.025} \tag{18}$$

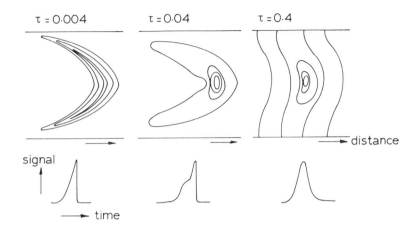

FIGURE 6   Calculated concentration profiles with corresponding response curves.   (From Ref. 15.)

$$\Delta t_B = \frac{35.4(2R)^2 f}{D^{0.36}} \left(\frac{L}{f_v}\right)^{0.64} \tag{19}$$

R and L are expressed in cm, f_v in mL min^{-1}, and D in cm^2 s^{-1}. f is a factor that has to be determined experimentally. Depending on the experimental conditions, the value of f lies between 0.5 and 1.0.

Recently, both Vanderslice et al. (15) and Wada et al. (16) have made simulation studies on the development of the concentration profile of a sample plug on passage through a straight tube; the latter authors have included the proceeding of a chemical reaction. Some of the results of Vanderslice et al. (15) are reproduced in Fig. 6, clearly demonstrating most of the aspects discussed above.

2.2 Tanks-in-Series Model

In chemical reaction engineering, flow patterns within a chemical reactor can be so complex that it is hardly possible to present even a less exact physical description of the system. Nevertheless, it can be of great value to obtain useful and transferable information about the flow behavior, even if physically meaningful details have got lost. Such a more macroscopic model is the tanks-in-series model. This model is based on the assumption that the fluid flows through a series of equal-sized, perfectly mixed tanks. It appears to be a one-parameter model, the only parameter being the total number of tanks. In this respect the result resembles the outcome of the axial-dispersed plug flow model, which is also a one-parameter model, in this case the axial Péclet number. As will be demonstrated later, both parameters can be interrelated.

The derivation of an equation of the tanks-in-series model starts with the mass balance equation for a component A around the ith tank of a series of n tanks

$$f_v(C_A)_i = f_v(C_A)_{i-1} - V_i \frac{d(C_A)_i}{dt} \tag{20}$$

where V_i is the volume of the ith tank. For equally sized tanks V_i can be replaced by V_{tot}/n, where n is the total number of tanks and V_{tot} is the total volume of the reactor. Solving this equation yields a recurrent relation that eventually leads to

$$C_A(\theta) = C_{A_0} \frac{n^n}{(n-1)!} \theta^{n-1} \exp(-n\theta) \tag{21}$$

in which $C_{A_0} = M/(V_{tot}/n)$ and $\theta = V_{tot}/f_v = t/t_v$. For large values of n (n > 5) the expression $(n!) \sim n^n e^{-n} \sqrt{2\pi n}$ applies and Eq. (21) can be rewritten as

$$C_A(\theta) = \frac{M}{V_{tot}/n} \sqrt{\frac{n}{2\pi}} \theta^{n-1} \exp[-n(\theta - 1)] \tag{22}$$

The variance can be calculated to be

$$\sigma_\theta^2 = \frac{1}{n} \tag{23}$$

$$\sigma_t^2 = \frac{V_{tot}^2}{f_v^2} \sigma_\theta^2 = \frac{V_{tot}^2}{n f_v^2} \tag{24}$$

For large values of n the calculated curves shown in Fig. 7 are quite similar to the ones obtained by means of the axial-dispersed plug flow model. So for n > 10 the variances can be used to correlate the number of tanks and the axial Péclet number

$$\sigma_\theta^2 = \frac{2}{Pe'_L} = \frac{1}{n} \quad \text{or} \quad n = \frac{Pe'_L}{2} \ (n > 10) \tag{25}$$

Of course, for a conduit to contain this minimum number of 10 hypothetical tanks, the conduit must be sufficiently long.

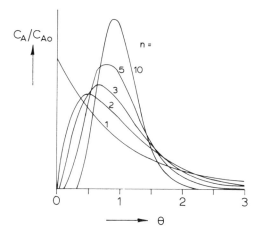

FIGURE 7 Response curves on perfect pulse injection for various numbers of tanks based on the tanks-in-series model.

2.3 Some Interim Conclusions

From theory it can be predicted that the transportation of a small injected sample plug will yield a more-or-less Gaussian-shaped response curve of small width provided that

$$\tau = \frac{t_v D}{R^2} = \frac{\pi R^2 L}{f_v} \frac{D}{R^2} = \frac{\pi L D}{f_v} > 0.5 \tag{26}$$

So L should be sufficiently long and the volumetric flow rate sufficiently low. If this condition holds true, a position is reached where the peak width is linear proportional to both the radius and the square root of the length, and inversely proportional to the volumetric flow rate. Of course, the opposite is true for the peak height since the area should remain constant.

The maximum sample frequency S_{max}, which is an important practical parameter, can be directly related to the standard deviation. To avoid excessive overlap between the two consecutive injected samples, the distance between the peaks at the end of the tubes should be approximately $6\sigma_t$, where σ_t is expressed in seconds. If we define S_{max} as the maximum number of samples that can be analyzed per hour, this leads to

$$S_{max} = \frac{3600}{6\sigma_t} = 600 \sqrt{\frac{24D}{t_v R^2}} \tag{27}$$

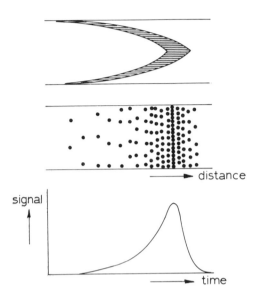

FIGURE 8 Two different spatial distributions of sample leading to
the same response curve.

To give an example, it is easy to calculate that for a residence time
of about 20 s, a tube radius of 0.3 mm, and a diffusion coefficient
of 2×10^{-9} m^2 s^{-1}, the value of S_{max} is approximately 170 samples
per hour. If the condition $\tau > 0.5$ is not fulfilled, tailing will occur
and the maximum sampling frequency will drop considerably. If we
use Vanderslice's formula (Eq. 19) for the same values of residence
time, the diameter and the diffusion coefficient S_{max} can be calculated to be about 30 samples per hour.

    At this stage it seems appropriate to emphasize that, in general,
it is very difficult, if not impossible, to relate the experimentally
obtained response curve in an unequivocal way to the actual spatial
distribution inside the system. This is due to the fact that most
detectors have some kind of averaging effect over the cross section
of conduits through which the sample is transported to the detector.
Different spatial distributions at the end of the reactor or mixing
conduit can lead to similar response curves. This is illustrated for
two simple cases in Fig. 8. Viewed in this light, one has to be
very careful in giving a physical explanation of the dispersion (factor) defined by Růžička and Hansen (17) as the ratio of the original
concentration in the injected sample (C°) to the maximum of the recorded curve (C_{max}). For instance, a large value of C°/C_{max} does
not necessarily mean a large dilution, as often suggested. This is

true only for a cross-sectional uniform distribution of the sample. Similarly, it is often taken for granted that the material is radially homogeneously dispersed when gradient techniques are used.

Furthermore, it has to be stressed that all the results discussed hitherto, however interesting they may be from a theoretical point of view, apply to straight tubes with a flow not obstructed by obstacles. In practice, tubes are often coiled, which causes secondary flow, and the various component parts of the system are connected by means of connectors that are not ideal and may contribute to radial mixing as well. Therefore, in experimental situations, Taylor-type flow may already occur at lower reduced times.

3 INFLUENCE OF VARIOUS PARTS OF FLOW INJECTION SYSTEM ON THE RESPONSE CURVE

In the preceding section, models for the mass transport were discussed for the basis component of any flow injection system (i.e., the transport, mixing, or reactor conduit). The dynamic behavior was presented by mathematical expressions describing the response to a perfect pulse or delta injection. Such expressions are therefore often called impulse response curves. However, the response or output signal of a system is also dependent on how the sample is injected as well as on the sample size, in other words, is dependent on the dynamic behavior of the injection system. Since any concentration input function, $C(t)$, can be thought to be composed of a sum of delta injections, the output from the conduit, $S_{out}(t)$, can be described as the convolution of this input function and the impulse response function (Fig. 9)

$$S_{out}(t) = C(t) * h_d(t)$$
$$= \int_0^\infty C(t - \tau) h_d(\tau) \, d\tau \qquad (28)$$

where τ is a dummy variable and $h_d(\tau)$ is the impulse response function. On convolution of the output signal of the conduit and the impulse response function of the detector, the ultimate response

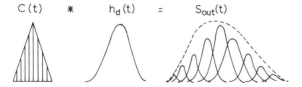

C(t)   * $h_d(t)$ = $S_{out}(t)$

FIGURE 9   Illustration of the process of convolution.

curve of the total system is obtained.   Under certain restricting
conditions it can be proven that the variances of the various individual functions are additive, leading to the following expression for
the total variance:

$$\sigma^2_{t,tot} = \sigma^2_{t,inj} + \sigma^2_{t,conduit} + \sigma^2_{t,det}$$

or, in a generalized form,

$$\sigma^2_{t,tot} = \Sigma_k \, \sigma^2_{\tau,k} \tag{29}$$

where k is a subsystem of the total system.   Successively, attention will be focused on the influence of the various subsystems,
such as the injection, detection, and other modules.

3.1   Influence of Injection Mode and Injection Volume

In most cases the sample will be introduced by means of an injection
valve, either with an external sample loop or with an internal bore
of the desired dimensions.   In the ideal situation insertion of the
sample volume in the flowing carrier stream occurs instantaneously
when the valve is switched, and a well-defined plug will be obtained.
Another possible type of injection, which is called the ideal time injection, can be approximately realized by using an injection syringe.
Here the sample is delivered to the flowing stream during a finite
time ΔT in such a way that the amount of sample introduced to each

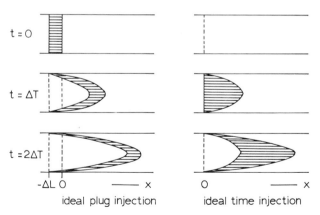

FIGURE 10   Distribution of sample in laminar flow without diffusion
for plug and "time" injection at times t = 0, t = ΔT, and t = 2 ΔT.

TABLE 2 Dependence of Mean Residence Time and Variance on Sample Volume for Various Injection Modes

	m	σ^2
Ideal plug injection	$1 + \dfrac{1}{2}\alpha$	$\dfrac{1}{n} + \dfrac{1}{12}\alpha^2$
Exponential injection	$1 + \alpha$	$\dfrac{1}{n} + \alpha^2$
Delta injection	1	$\dfrac{1}{n}$
Experimental results	$1 + 0.58\alpha$	$\dfrac{1}{n} + 0.105\alpha^2$ (short conduit)
		$\dfrac{1}{n} + 0.133\alpha^2$ (long conduit)

streamline is proportional to the velocity on that streamline. Both types of injection are depicted in Fig. 10 together with the sample distribution observed some time after the injection, assuming laminar flow conditions without taking diffusion into account (18). In a third type of injection it is assumed that the sample is contained in a single tank that is washed out by the carrier stream. In the ideal situation such a description leads to an exponential injection function.

Starting from a tanks-in-series model, Reijn et al. (18) were able to calculate the influence of the sample loop volume for some of the injection modes. The key parameter in the equations is the ratio (α) of the length of the sample loop and the length of the conduit, provided that loop and conduit have the same diameter

$$\alpha = \Delta L / L \tag{30}$$

The results with respect to the mean residence time (m) and the variance are summarized in Table 2.

For comparison, the corresponding values for a delta injection are included as well as experimentally obtained values for a commercially available injection valve provided with a pneumatic actuator. The results show that the input function for a properly designed injection valve corresponds to an almost ideal plug injection. The minor deviations can be attributed to the fact that on switching the valve it takes a small but finite time before the carrier stream can flow freely through the entire cross section into the sample loop.

3.2   Influence of Detection Mode

Two types of detectors can be distinguished: the *cup-mixing* and
the *mean-value* detectors.   In the cup-mixing type a flow rate
weighted-average value is measured

$$C_{cup} = \frac{\int_0^R C(r)v(r)2\pi r \; dr}{\langle v \rangle \pi R^2} \tag{31}$$

This situation is met in mass flow sensitive devices, such as continuous titrators and coulometric sensors as well as in cases where
fractions are collected and the mean concentrations of these fractions are determined.   Also, flame spectrometric detectors belong
to this category.   The mean-value detector "looks" through a plane
perpendicular to the direction of flow, and the measured concentration value is represented by

$$C_{mean} = \frac{\int_0^R C(r)2\pi r \; dr}{\pi R^2} \tag{32}$$

Examples of this type are the optical detector and the refractive
index detector.

From Eqs. (31) and (32) it is evident that the difference between the readout of both types of detectors vanishes if $v(r)$ approaches $\langle v \rangle$ (i.e., for low flow rates) and if the concentration
distribution entering the detector is independent of r [i.e., $C(r) =$
C = constant].   However, for a description of the influence of the
detector on the dynamic behavior of the whole system, it is not
sufficient to know what the detector is actually "seeing" but also
how fast it responds.   This aspect is related to the dynamic behavior of transducer and electronics and it is predictable from am-
plifier time constants (19).   A discussion of this contribution is
beyond the scope of this chapter.

3.3   Influence of Other Modules

Although flow injection systems in combination with flame spectroscopic detection techniques are generally used only for the fast and
reproducible introduction of samples into the spectrometer, separation and preconcentration modules can be incorporated in the manifold to increase selectivity (i.e., to eliminate an interfering matrix
from the analyte and/or to improve the sensitivity).   Both extraction systems (20) and gas diffusion membrane modules (21) have
been proposed for this purpose, the latter in combination with a
hydride generating system.   Such modules can contribute to the

overall sample dispersion to a considerable extent, so it would be
desirable to know their impulse response functions. Up to now, for
most such modules these functions are not available, but it has been
demonstrated that, for instance, the gas diffusion module can be
described adequately by a tanks-in-series model (22). A limiting
factor in the application of such modules in combination with an
atomic absorption spectrometer might be the minimum flow rate required by the nebulizer feed system. This aspect is discussed in
more detail in the following section.

4 SPECIFIC ASPECTS RELATED TO THE USE OF A FLAME ATOMIC ABSORPTION SPECTROMETER AS A DETECTOR

The usual flow-through detectors for flow injection methods are designed in such a way as to cause minimum disturbance of the flow
pattern. Such detectors seldom impose severe restrictions on the
input, apart, of course, from the concentration range. Atomic absorption spectrometric detectors are quite different in this respect.
Here factors such as flow rate are of such importance that it has
been stated that the properties of the atomizer dictate the design
and operation of the sample introduction system, especially for
liquid sample introduction with pneumatic nebulization (23,24). The
correctness of this statement will be better appreciated if it is realized how complex the whole signal generating process is. First,
the orderly sample flow is completely disrupted in the nebulizer,
and subsequently the aerosol, thus formed, has to be converted in
several steps to a population of free atoms. Due to the complexity,
it is difficult to present a unified theory and to give a derivation
of a generally applicable equation for the dynamic properties of the
detection system in terms of physical and geometrical parameters.
   Apart from Tyson's work (25), hardly any information is available in the literature on the transient behavior. This is not surprising because in "classical" flame atomic absorption spectrometry
steady-state signals are used almost exclusively, with nebulization
efficiency and detection limits as the main objects of interest.

4.1 Sample Introduction in Nebulizer

At present three methods of sample introduction have been proposed
for flow injection atomic absorption spectrometry. Fukamachi and
Ishibashi (26) injected the sample into a carrier stream propelled
solely by the aspirating action of the nebulizer. Their approach
looks very attractive because of its simplicity, but it is not the optimum solution in terms of sensitivity and reproducibility. In a procedure suggested by Wolf and Stewart (27), the carrier stream is

pumped directly into the nebulizer. They found that the nebulization efficiency increases with a decrease in sample flow rate. A
small but distinct effect was observed on variation of the natural
aspiration flow rate. This is the flow rate at which the solution
would be aspirated when the pump is removed; it can be adjusted
by varying the velocity of the nebulizer gas past the inlet needle
and by the position of this needle. The effect of pumping and aspiration rate was explored further by Brown and Růžička (28).

Their results indicate that for their system using a barrel nebulizer optimum performance is achieved by a pumping rate that is
about twice the natural aspiration rate. However, this asks for a
relatively high pumping rate, whereas in terms of reagent economy—
if such reagents are used in the system—lower rates would be desirable. They suggest that a solution to this problem might be to
boost the flow rate just before the nebulizer by the confluence of
an additional stream. It is true that this will lead to dilution, but
the negative effect on the signal was found to be offset by the enhanced sensitivity due to the increased flow rate in the nebulizer.
So in the end no adverse effect was observed on the total response.
Also, Yoza et al. (29) stress this point that flow rate to the burner
should not be less than the aspiration rate because otherwise, erratic results will be obtained due to air entering the injection system through the connectors. They propose a "compensation method"
in which the sample plug is transported by pumping of the carrier
stream, but where the additional amount of solvent is aspirated as
shown in Fig. 11. The coil C_1 is meant to damp the shock during
injection, and C_2 is a mixing coil which was found to improve the
reproducibility.

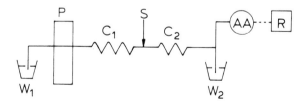

FIGURE 11 Manifold for FIA-AAS with compensation. AA, atomic
absorption spectrometer; R, recorder; S, sample injector; P, pump;
C_1 and C_2, coils; W_1, carrier solution reservoir; W_2, water reservoir.

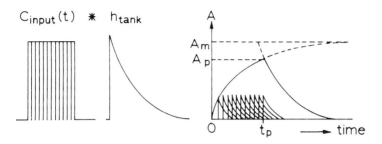

FIGURE 12   Absorbance-time profile for plug injection in a single tank.

4.2   Dynamic Model for an Atomic Absorption Spectrometer

The following discussion is based mainly on the work of Tyson and co-workers (25,30). They have shown that the response of an atomic absorption spectrometer to a stepwise change in concentration from zero to C_{inj} can be approximately described by an exponential relation

$$A = kC_{inj}[1 - \exp(-f_v t/V)] \tag{33}$$

where A is the absorbance and V is a constant with the dimension of a volume. The response resembles the one obtained on the passage of a concentration step through a single well-stirred tank of volume V. The volume of this hypothetical tank was found to be on the order of 40 to 100 µL for an optimum flow rate. Such stepwise concentration changes are obtained on starting the aspiration from a large sample volume. In flow injection, however, the sample volume is limited and the loss of material from the tank will start to prevail as soon as the end of the sample plug has reached the tank, leading to an absorbance-time profile as depicted in Fig. 12. In fact, this result is obtained on convolution of a rectangular block input concentration profile with the impulse response function of a single tank

$$A = const(C_{input} * h_{tank}) \tag{34}$$

where

$$C_{input} = C_{inj}(0 < t < t + \Delta T)$$

and

$$C_{input} = 0(t < 0; \; t > t + \Delta T)$$

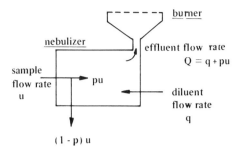

FIGURE 13   Schematic representation of extended-tank model.

ΔT is the duration of the input or $\Delta T = \Delta V/f_v$, where ΔV is the injection volume.   Using Eq. (21) with $n = 1$ yields

$$h_{tank} = \exp(-t/t_v) = \exp(-f_v t/V)$$

The maximum absorbance is observed at $t_p = \Delta T = \Delta V/f_v$ and can be represented by

$$A_p = A_m[1 - \exp(-f_v t_p/V)] = A_m[1 - \exp(-\Delta V/V)] \tag{35}$$

Although this simplified approach correctly accounts for the effect of the sample volume on the peak height, it suggests an independence of the flow rate which is incorrect.   Therefore, an extension of the single-tank model was proposed as shown in Fig. 13.   The value p represents the fraction of the sample stream that is converted to aerosol that will reach the burner.   The mass balance for the analyte reads

$$\left(\frac{dm}{dt}\right)_{tank} = \left(\frac{dm}{dt}\right)_{in} - \left(\frac{dm}{dt}\right)_{out} = pf_v C_{inj} - QC \tag{36a}$$

or on division by V,

$$\frac{dC}{dt} = \frac{pf_v C_{inj}}{V} - \frac{QC}{V} \tag{36b}$$

To get the response to a stepwise concentration change from 0 to C_{inj} at a steady flow rate of f_v, Eq. (36b) has to be integrated. If we use the initial condition $C_{inj} = 0$ at $t = 0$, the following solution is found:

$$C = \frac{pf_v C_{inj}}{Q} [1 - \exp(-Qt/V)] \tag{37}$$

If it is assumed that the detector response (absorbance) is proportional to the rate at which the analyte enters the burner (i.e., $A = kQC$), Eq. (37) can be written as

$$A = kpf_v C_{inj}[1 - exp(-Qt/V)] \tag{38}$$

It has to be noticed that the response time is determined now by the effluent flow rate through the nebulizer, not by the sample flow rate alone: Increase of the diluent flow rate will yield a faster response. The signal height will be affected as well, not only by means of the variables Q and f_v but also because of the flow dependence of p. Comparison of experimentally obtained results with the theoretically predicted ones shows that the extended tank model gives at least qualitatively a correct description of the influence of the various parameters on the behavior of the detector.

5 DIMENSIONAL ASPECTS OF FLOW INJECTION SYSTEMS

A novel approach for constructing flow injection systems is the so-called "integrated microconduit" (31,32). In this approach, parallel to the development of concepts in integrated electronic circuitry, most of the functional parts of the manifold are combined in a small single block. In the examples presented in the literature up to now, the grooves were manufactured by mechanical engraving procedures and the sizes of the conduits were not reduced significantly, but present lithographic techniques offer the possibility of obtaining manifolds of much smaller dimensions. The enhanced reproducibility that is observed experimentally together with a strongly reduced reagent and sample consumption warrants the expectation that work along these lines will be continued. Therefore, it seems appropriate to pay some attention to the effects of miniaturization from a theoretical point of view (33). The set of equations on which the discussion of scaling down will be based is summarized in Table 3.

Apart from the pressure drop equation, all the others have been introduced in the preceding sections. The effect of scaling down (i.e., reduction of the conduit dimensions) can be treated in several ways. Two possible approaches are chosen. At first it is assumed that

1. The residence time should remain constant in order to maintain the same degree of conversion for those cases where a chemical reaction has to proceed.
2. The pressure drop across the conduit should not increase to avoid problems with the pump and the tightness of the connections.

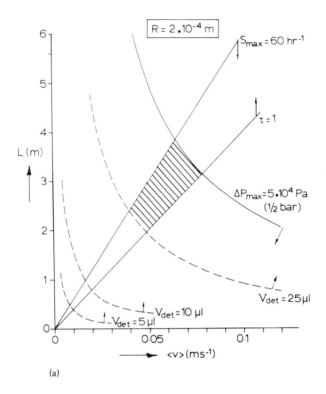

FIGURE 14   Graphs showing the constraints discussed in the text:
(a) R = 2 × 10⁻⁴ m; (b) R = 10⁻⁴ m. The enclosed hatched areas
indicate allowable L-⟨v⟩ combinations so that $S_{max} > 60$ h⁻¹; $\tau > 1$;
$V_{det} > 25$ μL, 10 μL, etc.; and $\Delta P < 1/2$ bar.

From Table 3 it can be seen that for these two conditions to
become fulfilled, an x-fold decrease in tube radius should be accompanied by a decrease of the product ⟨v⟩L with a factor x^2 in order
to keep ΔP constant. Because $t_V = L/\langle v \rangle$ should also remain constant, this means that both L and ⟨v⟩ have to decrease by a factor
x. This results in a reduction of both $Pe_R = Re \cdot Sc$ and τ by a
factor x^2, which implies that the assumption of Taylor-type flow
will, in general, be more appropriate. However, technical difficulties may arise because the corresponding decrease in conduit volume
by a factor x^3 should be accompanied by a similar decrease in injection volume, whereas the detector volume, which is related to σ_V,

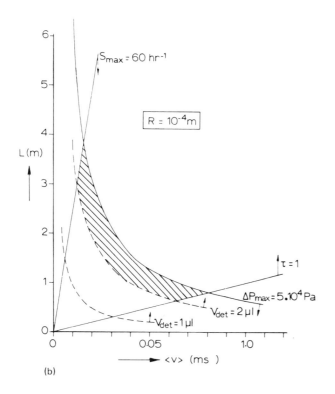

(b)

TABLE 3 Summary of Relations to Be Used in the Discussion of Effects of Scaling Down

$$\Delta P = \frac{8\eta \langle v \rangle L}{R^2}$$

$$t_v = \frac{L}{\langle v \rangle}$$

$$\tau = \frac{Dt_v}{R^2}$$

$$Re \cdot Sc = \frac{2\langle v \rangle R}{D}$$

$$\sigma_t = \sqrt{\frac{t_v R^2}{24D}}$$

$$V_{conduit} = \pi R^2 L$$

$$f_v = \langle v \rangle \pi R^2$$

$$\sigma_v = f_v \sigma_t$$

$$S_{max} = \frac{600}{\sigma_t}$$

TABLE 4 Constraints and Corresponding L versus ⟨v⟩ Relations

Constraint	Relation	Remarks
$\tau \geq 1$	$L \geq \dfrac{R^2}{D} \langle v \rangle$	
$S_{max} \geq 60 \text{ hr}^{-1}$	$L \leq \left(\dfrac{600}{S_{max}}\right)^2 \dfrac{24D\langle v \rangle}{R^2}$	
$\Delta P \leq \Delta P_{max}$	$L \leq \dfrac{P_{max}R^2}{8\eta} \dfrac{1}{\langle v \rangle}$	
$V_{det} \geq V_{det,min}$	$L \geq \dfrac{96DV^2_{det,min}}{\pi^2 R^6} \dfrac{1}{\langle v \rangle}$	$V_{det} = \dfrac{1}{2}\sigma_v$

should be reduced even by a factor x^4. It is obvious that especially with the detector, one will run very quickly into technological problems.

Another approach in discussing the effect of scaling is to introduce some constraints and to evaluate the domain where the constraints are not surpassed. For the graphical representation L versus ⟨v⟩ plots are chosen [Figs. 14(a) and 14(b)] because these two variables are most easily adapted in flow injection manifolds. Consequently, the constraints have to be expressed in terms of L and ⟨v⟩ to enable the introduction in the graphs. Table 4 presents a summary of the corresponding relations.

The hatched areas in Figs. 14(a) and 14(b) show the domains where the conditions are met. The constraint value $\Delta P_{max} = 5 \times 10^4$ Pa (~0.5 bar) is chosen because it is supposed that peristaltic pumps are used. It is obvious that a reduction of the tube radius by a factor of 2 already shows a considerable change in the picture: The slope of the S_{max} line strongly increases, whereas the slope of the τ-limitation line decreases; at the same time the ΔP line shifts in the direction of the origin, whereas the V_{det} line shifts in the opposite direction. This latter shift especially illustrates once again the severe demands put on the detector volume.

Commercially available atomic absorption spectrometers used as a detector for flow injection analysis exhibit an effective volume of such size (>40 µL) that it is not worthwhile to reduce the dimensions of the manifold conduits as they are normally used nowadays. If, however, miniaturization of flow injection atomic absorption spectrometry is pursued, for example, for reasons of sample and reagent economy, attention will have to be focused on diminution of the peak-broadening effects of the nebulizer system.

6 CONCLUDING REMARKS

In this chapter some theoretical concepts currently used in the literature on flow injection analysis have been discussed. Although no generally valid analytical solution is presented for the equation describing the process of sample dispersion in flowing media, useful relations can be derived when simplifying assumptions are made. The relations thus obtained provide at least qualitative information about the performance of manifolds with respect to the variation of parameter values such as flow rate, tube radius, tube length, and so on, and help the analyst in developing guidelines for setting up and optimizing flow injection manifolds. The versatility of the technique is outlined in the following chapters. The broad field of application as far as flow injection atomic absorption is concerned is based on three advantages:

1. High sample throughput
2. Small sample volumes, normally < 0.5 mL
3. Improved precision and sensitivity

REFERENCES

1. O. Levenspiel, *Chemical Reaction Engineering*, 2nd ed., Wiley, New York, 1972.
2. G. F. Froment and K. B. Bischoff, *Chemical Reactor Analysis and Design*, Wiley, New York, 1979.
3. W. J. Beek and K. M. K. Muttzall, *Transport Phenomena*, Wiley, New York, 1975.
4. R. Tijssen, Axial dispersion and flow phenomena in helically coiled tubular reactors for flow analysis and chromatography, *Anal. Chim. Acta*, *114*:71 (1980).
5. W. E. van der Linden, Miniaturisation in flow injection analysis; practical limitations from a theoretical point of view, *TrAC*, *6*:37 (1987).
6. O. Levenspiel and K. B. Bischoff, Patterns of flow in chemical process vessels, *Adv. Chem. Eng.*, *4*:95 (1963).
7. G. I. Taylor, Dispersion of soluble matter in solvent flowing slowly through a tube, *Proc. Roy. Soc. (London) A.*, *219*:186 (1953).
8. M. J. Lighthill, Initial development of diffusion in Poiseuille flow, *J. Inst. Math. Its Appl.*, *2*:97 (1966).
9. H. R. Bayley and W. B. Gogarty, Numerical and experimental results on the dispersion of a solute in a fluid in laminar flow through a tube, *Proc. Roy. Soc. (London) A.*, *269*:352 (1962).
10. V. Ananthakrishnan, W. N. Gill, and A. J. Barduhn, Laminar dispersion in capillaries. Part 1. Mathematical Analysis, *AIChEJ.*, *11*:1663 (1965).

11. R. Aris, On the dispersion of a solute in a fluid flowing through a tube, *Proc. Roy. Soc. (London) A.*, *235:*67 (1956).

12. H. Bate, S. Rowlands, J. A. Sirs, and H. W. Thomas, The dispersion of diffusible ions in fluid flow through a cylindrical tube, *Br. J. Appl. Phys.*, *2:*1447 (1969).

13. H. Bate, S. Rowlands, and J. A. Sirs, Influence of diffusion on dispersion of indicators in blood flow, *J. Appl. Physiol.*, *34:*866 (1973).

14. J. T. Vanderslice, K. K. Stewart, A. G. Rosenfeld, and D. J. Higgs, Laminar dispersion in flow-injection analysis, *Talanta, 28:*11 (1981).

15. J. T. Vanderslice, A. G. Rosenfeld, and G. R. Beecher, Laminar-flow bolus shapes in flow injection analysis, *Anal. Chim. Acta, 179:*119 (1986).

16. H. Wada, S. Hiraoka, A. Yuchi, and G. Nakagawa, Sample dispersion with chemical reaction in a flow-injection system, *Anal. Chim. Acta, 179:*181 (1986).

17. J. Růžička and E. H. Hansen, *Flow Injection Analysis*, Wiley, New York, 1981.

18. J. M. Reijn, W. E. van der Linden, and H. Poppe, Some theoretical aspects of flow injection analysis, *Anal. Chim. Acta, 114:*105 (1980).

19. H. Poppe, The performance of some liquid phase flow-through detectors, *Anal. Chim. Acta, 145:*17 (1983).

20. L. Nord and B. Karlberg, Sample preconcentration by continuous flow extraction with a flow injection atomic absorption detection system, *Anal. Chim. Acta, 145:*151 (1983).

21. G. E. Pacey, M. R. Straka, and J. R. Gord, Dual phase gas diffusion flow injection analysis/hydride generation atomic absorption spectrometry, *Anal. Chem.*, *58:*502 (1986).

22. W. E. van der Linden, Membrane separation in flow injection analysis; gas diffusion, *Anal. Chim. Acta, 151:*359 (1983).

23. R. F. Browner and A. W. Boorn, Sample introduction; the Achilles' heel of atomic spectroscopy? *Anal. Chem.*, *56:*787A (1984).

24. R. F. Browner and A. W. Boorn, Sample introduction techniques for atomic spectroscopy, *Anal. Chem.*, *56:*875A (1984).

25. J. M. H. Appleton and J. F. Tyson, Flow injection atomic absorption spectrometry: The kinetics of instrument response, *J. Anal. At. Spectrom.*, *1:*63 (1986).

26. K. Fukamachi and N. Ishibashi, Flow injection-atomic absorption spectrometry with organic solvents, *Anal. Chim. Acta, 119:*383 (1980).

27. W. R. Wolf and K. K. Stewart, Automated multiple flow injection analysis for flame atomic absorption spectrometry, *Anal. Chem.*, *51:*1201 (1979).

28. M. W. Brown and J. Růžička, Parameters affecting sensitivity and precision in the combination of flow injection analysis with flame atomic absorption spectrophotometry, *Analyst, 109:*1091 (1984).

29. N. Yoza, Y. Aoyagi, S. Ohashi, and A. Tateda, Flow injection system for atomic absorption spectrometry, *Anal. Chim. Acta, 111:*163 (1979).

30. J. F. Tyson, J. M. H. Appleton, and A. B. Idris, Flow injection calibration methods for atomic absorption spectrometry, *Anal. Chim. Acta, 145:*159 (1983).

31. J. Růžička, Flow injection analysis; from test tube to integrated microconduits, *Anal. Chem., 55:*1040A (1983).

32. J. Růžička and E. H. Hansen, Integrated microconduits for flow injection analysis, *Anal. Chim. Acta, 161:*1 (1984).

33. W. E. van der Linden, Flow injection analysis—the intermarriage of analytical chemistry and chemical engineering science, *Anal. Chim. Acta, 180:*20 (1986).

3

Basic Components and Automation

JACOBUS F. VAN STADEN *Department of Chemistry,*
University of Pretoria, Pretoria, South Africa

1 INTRODUCTION

The use of flow injection analysis (FIA) techniques for the preparation of solutions prior to their aspiration into atomic spectroscopic systems forms the basis of interfaced flow injection/atomic spectroscopy. Although FIA has become a well-known name to most chemists over the past decade, the real success of practical instrumental FIA analyzers in routine analytical laboratories depends on the type of basic components used and the construction thereof into a working FIA system.

In most modern spectrometric instruments a high rate of sample detection is usually possible, but in many applications this rate is not easily matched by the preparation steps needed to obtain a solution or reaction product suitable for detection. In this regard FIA offers several advantages. Among the advantages are

High sample throughput; 200 to 300 samples per hour
Shorter startup times (no bubble trouble); except for
        detector warm-up, the system is ready for instant operation
        immediately after the sample is injected
Reduced residence times; readout time is about 3 to 40 s
Shorter reaction times (3 to 60 s)
Easy switching from one analysis to another; manifolds are easily
        assembled and/or exchanged
Reliability
Reproducibility
Low carryover
High degree of flexibility in terms of sensitivity (greater variety
        of sample volumes and concentration ranges is possible)
Low reagent consumption
Ease of automation

However, the real advantages to the spectroscopist are based on a combination of the following three principles (1):

1. Reproducible sample injection
2. Controlled precise dispersion of the injected sample zone
3. Reproducible timing of the movement of the dispersed sample
    zone through the manifold system to the detection device

These objectives are possible only if a well-defined sample plug is injected into a precise continuous flow carrier stream in such a way that the hydrodynamic flow conditions are not altered. Furthermore, the sample plug must be processed in such a manner that a smooth hydrodynamic flow system in the manifold and detector system is obtained.

FIGURE 1 Tecator model 5020 flow injection analyzer with two independently adjustable peristaltic pumps, each with four channels. Thumbwheels are used for setting and stop times for both pumps as well as for setting the injection time and selecting the evaluation mode. A thermostatted compartment houses the chemical manifold, the injection valve, and the optional detection. The instrument is also equipped with a keyboard for entry of standard concentration values, number of injections per sample, and type of calibration graph (linear or nonlinear), as well as a display for results and function codes. (Reproduced with permission of Tecator AB, Höganäs, Sweden.)

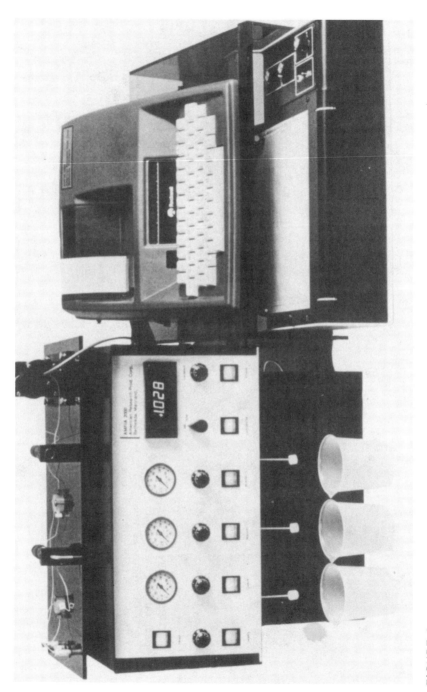

FIGURE 2

During its conception in the mid-1970s homemade FIA systems were used by the early groups (Růžička and Hansen in Denmark; Stewart and co-workers in the United States; Bergamin, Zagatto, and Krug in Brazil; Karlberg in Sweden; and Basson and van Staden in South Africa) to trigger off the technique. The components from which the homemade systems were constructed vary from LEGO toy building blocks in Denmark to segmented flow analyzer parts and high-performance liquid chromatography components in South Africa and the United States. As the concept jumps into analytical reality, commercial equipment (see Figs. 1 to 3) becomes available. Most of the commercial equipment is nowadays slotted into routine analytical laboratories as dedicated FIA analyzers.

2 SAMPLING SYSTEMS

Sampling forms a crucial and indispensible part of a flow injection analyzer. The importance of this part of the analyzer is spelled out in part of the name coined by the originators of this concept: namely, injection. Automation of sampling in flow injection analysis can be divided into the following elements: (a) a sampling system and (b) a liquid delivery system. Flow injection of a discrete volume of sample occurs where solutions in these two flow systems meet.

The entire sampling system is a combination of an automatic sampler and a sample injector. Samples are drawn (or pumped) from an automatic sampler or directly from the sample source (in industrial process control) to the sample injector. The amount of sample drawn (or pumped) from a sample cup is usually more than the amount injected into the carrier stream as a plug due to sample washing in the sample loop. The first amount is also regulated by the inner bore of the pumping tube used, the rotation rate of the peristaltic pump, and the time of the sample needle inside the sample cup.

FIGURE 2 AMFIA-2000 flow injection analysis system with an electromechanical microprocessor control 316 stainless steel; seal carbon fiber injection valve (sample control left). Variable-speed motor valveless metering-type pumps with adjustable volumes are used to supply one carrier stream and two reagent streams (if necessary). The chemistry module is temperature controlled. The system is also equipped with a digital readout and is coupled to a microcomputer. (Reproduced with permission of American Research Products Corp., Kensington, Maryland.)

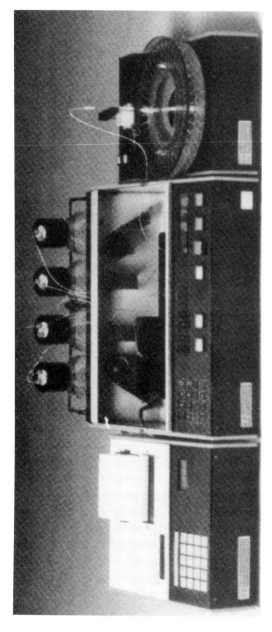

FIGURE 3 Skalar (Cenco, Breda Scientific) model SA-20 flow injection analyzer with a 12-channel peristaltic pump rotating usually at 10 rpm (20 rpm or 30 rpm). The 5101 chemistry module is temperature controlled and is also equipped with an autosampler, injection valve, and an optional detector. The entire system is coupled to a recorder and data system. (Reproduced with permission of Skalar Analytical, Breda, Holland.)

FIGURE 4 Tecator model 5007 sampler. (Reproduced with permission of Tecator AB, Höganäs, Sweden.)

FIGURE 5   Skalar model 1000 sampler.   (Reproduced with permission of Skalar Analytical, Breda, Holland.)

2.1 Samplers

A variety of autosamplers are available.   The autosampler normally consists of a turntable holding up to 40 sample vessels.   Sample volumes can be dispensed with this apparatus, allowing both single and multiple analysis to be made.   After each sample insertion, the sampling probe is rinsed with distilled water.   The Tecator model 5007 sampler (Fig. 4) has a set of forty 8.5-mL sample cups and is designed for use with the Tecator 5020 flow injection analyzer, from where sampling and wash cycles are regulated.   Other features of the sampler are injection times from 0 to 90 s in 10-s steps, or the option of injection length (IL, sample length) by using the thumbwheels on the 5020 analyzer, multiple injections per sample via the keyboard of the 5020 unit, as well as a preadjusted sample cup number control.   In the standard Skalar model 1000 sampler (Fig. 5) the sampling control keyboard is on the sampler unit.   This sampler is equipped with the following features:

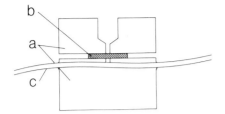

FIGURE 6   Septum injection device.   a, Perspex blocks; b, silicone
rubber septum; c, polyethylene tubing.   [Reproduced with permission of Elsevier Science Publishing (2).]

Sample cups capacity: 2 × 40 cups (1 × 40 cups) with a volume of
      4 mL each or 1 × 40 cups with a volume of 8 mL each
Preadjusted sample cup number control with alarm
Sampling time adjustable from 0 to 999 s in 1-s steps
Wash time adjustable from 0 to 999 s in 1-s steps
Small peristaltic pump in the wash cycle as an integral part of the
      sampler
Duplicate sample taken from 1 cup

2.2   Valves

The main purpose of sample injection is to insert a well-defined
discrete plug or zone of aqueous sample solution into a continuously moving carrier stream in such a way that the movement of the
stream is not disturbed.   The amount of sample need not be known
accurately, but it must be introduced into the carrier stream precisely, so that the volume and length of the plug can be produced
exactly from time to time.   Furthermore, the sample volume injected
is normally smaller than the sample volume required for the usual
continuous sample delivery.

In the early stages of development in FIA systems, an aqueous
sample was introduced manually (2) by a hypodermic syringe through
a septum device (Fig. 6) into a continuously moving unsegmented
carrier stream of water or reagent solution, an injection principle
which was also adapted to flow injection/atomic spectroscopy systems (3-6) and flow injection-inductively coupled atomic emission
spectroscopy (7).   With this sluice-type injector (time injection)
the sample is introduced as a flow increment during a finite period
into the carrier stream flow by a syringe.   This is done by introducing a hypodermic syringe needle through a self-sealing rubber
septum and injecting measured volumes from an attached syringe,
usually at low pressures.   The rubber septum, which is part of
this injector, may not be compatible with all samples and carrier

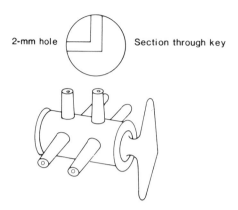

FIGURE 7   Stopcock injection valve.   [Reproduced with permission of the Association of Clinical Biochemistry (9).]

stream solutions.   Repeated injections with this type of device will often result in some of the elastomer being exposed to the carrier stream solution or industrial sample, resulting in chemical attack with possible chemical leaching.   Components leaching from the septum material may adversely affect the final results obtained.   One of the main disadvantages of a septum is that leakage may occur after repeated injections on the same spot, and pieces of the septum eventually become detached and block the flow line (8).   The use of a septumless injector eliminates these problems and even allows for reliable operation at high pressures.

The duration and magnitude of the temporary flow change, caused by the sample injection, are dependent on the time required to push the plunger into the syringe.   Uneven movement of the plunger can result in additional sample dispersion and peak asymmetry, and because of this disadvantage, the use of sluice-type injectors is decreasing.

It is also possible to replace a slice of carrier stream instantaneously by an identical plug (or slug) of the sample in an ideal plug injection using a loop-type injector.   In all loop injectors the sample is first introduced (injected, drawn, or pumped) into a predetermined volume from the sample source (turntable or directly in on-line monitoring).   This may either be the internal volume of the valve port or an external sample loop.   To inject, the sample volume in the loop is diverted into the carrier stream by a valve(s).   With this movement the sample is introduced as a well-defined cylindrical slug (plug, zone) into the carrier stream.   At the moment of introduction, the sample slug is accelerated by the carrier stream until it reaches the same flow rate as the carrier stream.

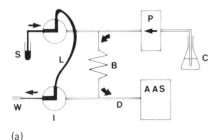

(a)

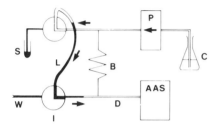

(b)

FIGURE 8 Stopcock injection valve with FIA-AAS system. A pump,
P, continuously propels carrier solution, C, through the system.
(a) The stopcock injection valve, I, in the fill position. Excess
sample, S, going to waste, W. In this mode carrier solution flows
through the bypass tube, B. (b) The valve in the analyze position. Carrier solution sweeps the sample from the valve through
the dispersion tube, D, and into the atomic absorption spectrometer,
AAS. [Reproduced with permission of the Association of Clinical
Biochemistry (9).]

This leads to physical mixing in the sample/carrier stream interface
zones with sample dispersion as well as chemical interface reactions.
    Various models of automatic injection systems have been de-
scribed in FIA, ranging from stopcocks (9), rotary (10-15), slider
valves (16,17), solenoid (18), commercially available valves, like
Carle (19-22), Valco (23,24), Perfektum (25,26), Rheodyne (27-33),
Altex (34), and Tecator (35,36), to the double proportional injector
of Bergamin et al. (37-48). Although some of these valves have
not been adapted exclusively to systems where atomic spectroscopy
fulfills the role of the detector, they appear to be suitable for FIA
work and are therefore included.
    Rocks and co-workers (9) introduced an inexpensive double
three-way stopcock valve (Fig. 7) into a FIA-AAS system with the
following illustrative operation (Fig. 8). The valve is equipped

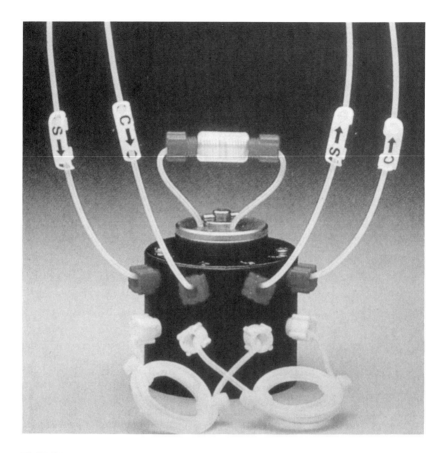

FIGURE 9   Tecator patented rotary valve furnished with a bypass.
(Reproduced with permission of Tecator AB, Höganäs, Sweden.)

with 0.7-mm narrow-bore PTFE tubing as sample loop with variable
but fixed length L.   The system also contains a bypass tube, B,
of higher hydrodynamic resistance than the valve to allow the carrier flow to continue when the valve loop is filled [Fig. 8(a)].   To
introduce the sample into the carrier stream, the valve is turned
[Fig. 8(b)] and the sample is swept into the analytical conduits of
the FIA-AAS system.

Most workers have developed and preferred to use rotary
valves; custom-built valves of this type have been described by
Růžička and Hansen (10,11,13,14), Anderson (12), Jørgensen and
Regitano (15), and Mindegaard (49).   The rotary sliding valve described by Růžička and Hansen (10,11,13,14) and adapted to the

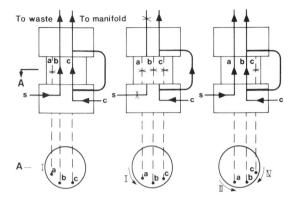

FIGURE 10 Operating principle for the Tecator injection valve.
[Reproduced with permission of International Scientific Communications, Inc. (35).]

Tecator model (35) consists of three layers, where a center movable
Teflon rotor is sandwiched between two stators of plexiglass or PVDF
(Fig. 9). The concept on which the valve operates is as follows
(Fig. 10). In the filling position (I) the sample streams fill the
bore b and the carrier stream goes through bore c to the manifold.
The center piece is then turned (II). During the turn all bores are
inaccessible to any of the streams. The carrier stream continues unobstructed as it now flows through the bypass. In the inject position (III) the sample in bore b is inserted into the carrier stream.
The flow in the bypass ceases, as the hydrodynamic resistance is
high due to its small diameter. Finally, the center piece is returned
to its filling position (IV).
      The valve consists of two identical injection parts, so that two
samples, or one sample and one reagent, can be injected simultaneously. A two-layer valve with a Teflon rotor mounted on one side
of a PVC stator block was also described (50,51). According to
Jørgensen et al. (51), the stators and rotors must have a number
of perfectly aligned bores to allow free passage of sample and reagent stream without any change in the hydrodynamic flow conditions
of the system and for defining sample size. Furthermore, the rotor
must be switchable between two exactly defined positions. If switching is performed slowly, each channel must be furnished with a bypass line, or pumping must be stopped during switching, as in the
commercial FIA-5020 system, to avoid pressure buildup in the pumping tubes.
      Stewart et al. (16) used a stream sampling valve consisting of
two four-way slider valves with two pneumatic actuators (activated

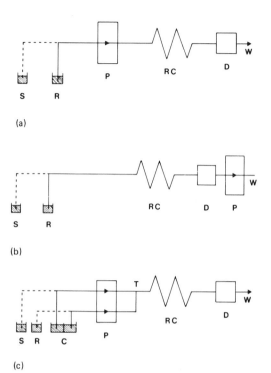

FIGURE 11 Valveless injection system for controlled dispersion apparatus. The pump, P, driven by a stepping motor under microcomputer control, propels fluids through the reaction coil, RC, through the photometric detector, D, and to waste, W. (a) The probe moves from reagent, R, to sample cup, S, and the pump rotates through several degrees, aspirating a small quantity of sample. The probe is then returned to the reagent reservoir and the pump restarts, propelling the sample through the reaction coil, where it mixes with the reagent to produce a color. (b) In this arrangement the pump is placed downstream of the detector. (c) Sample and reagent are simultaneously aspirated into independent carrier streams, C, of deionized water. The sample and reagent zones merge at T. [Reproduced with permission of the American Association for Clinical Chemistry (56).]

by compressed air) and a depulsing system, which is also the case in the pneumatically actuated slide valve system of Baadenhuijsen and Seuren-Jacobs (52). Mottola et al. (25,26) used a two-way automatic valve following the design of a Perfektum valve in a forced-flow injection system. Bergamin et al. (39) have initiated the double proportional sliding valve injector-commutator (37-48), which leads to the well-known merging-zone principle, in which the sample and reagent solutions are separately introduced into carrier streams (e.g., pure water) such that the sampling zone meets the reagent solution zone in a controlled manner. This can be achieved in two different ways: by intermittent pumping (53) or by the use of a multiple injection (39,49,54).

The potential of merging-zone-sampling processes has been discussed by Reis et al. (48), who also showed the usefulness of zone sampling in routine work. Basson and van Staden (19-22,55) used the Carle valve system and also described the use of a valve minder system (19,55) which enabled them to run more than one channel with the same valve minder system. A timed solenoid valve injector which can provide samples in a continuous range of sizes at any time intervals has also been described by Rothwell and Woolf (18). The various injection parts and valves used in liquid chromatography and available in many laboratories can also be used in FIA systems. If the reagent is corrosive and likely to attack the materials of which the valve is constructed, water should be used as the carrier stream.

Rocks and co-workers (56,57) designed a valveless injection system which consumes minimum amounts of sample and reagent. By using a computer-controlled peristaltic pump and aspiration probe a sample is aspirated into a probe or pump tube by the precise rotation of a peristaltic pump driven by a stepper motor (Fig. 11). The probe normally rests in the reagent solution. To sample, the pump is stopped and the probe is transferred to the sample container. The pump is then activated for a predetermined angular movement to draw up a precise volume of sample and then stopped. The probe is returned to the reagent carrier solution and the pump is again restarted, propelling the sample through the manifold system to the detector. In a more advanced version the same approach is combined with the merging-zone technique.

With the hydrodynamic injection technique described by Ružička and Hansen (58), part of a carrier stream tubing is used as a sample loop (Fig. 12). In the sampling position the flow rate of the carrier stream is stopped and the sample is pumped into a geometrically well-defined conduit of carrier stream tube of length L and internal radius r. In the injection mode the sample stream is stopped and the sample volume, $S_V = \pi r^2 L$, is subsequently propelled downstream by the carrier stream. By using an alternate and intermittent pumping of carrier and sample solutions, it was possible to form a well-defined sample zone which was inserted into the carrier stream. Two peristaltic pumps were used to control the movement of sample and carrier solutions (Fig. 13). A list of some automatic injection systems is given in Table 1.

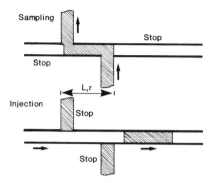

FIGURE 12 Principle of hydrodynamic injection. [Reproduced with permission of Elsevier Science Publishing (58).]

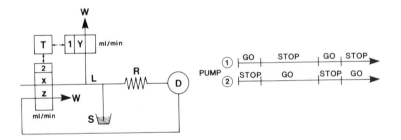

FIGURE 13 Manifold for hydrodynamic injection. The sample solution is aspirated by pump 1, operating at a pumping rate of y mL min^{-1}, and a fixed volume of sample from reservoir S passes into conduit L. Subsequently, pump 2 is activated, pumping at rates x = z mL min^{-1}, and the sample is flushed through reactor, R, to the detector, D. The operation of the two pumps is controlled by the timer, T, the time sequence of events being as depicted. [Reproduced with permission of Elsevier Science Publishing (58).]

TABLE 1 List of Some Automatic Injection Systems

Authors	System	References
Růžička and Hansen	Rotary sliding valve consisting of three layers; movable Teflon rotor sandwiched between two stators	10, 11, 13, 14
Anderson	Rotary valve (same principle as above)	12
Jørgensen and Regitano	Rotary valve (same principle as above)	15
Mindegaard	Rotary valve (same principle as above)	49
Jørgensen et al.	Versatile pneumatically operated two-layer rotary valve for diversion of analytical streams	51
Mottola et al.	Two-way automatic valve following the design of a Perfektum valve	25, 26
Stewart et al.	Two four-way slider valves with two pneumatic actuators (activated by compressed air)	16
Baadenhuijsen and Seuren-Jacobs	Same type as above	52
Bergamin et al.	Have initiated the double proportional sliding valve injector-commutator, which leads to the well-known merging zone principle	37–48
Rothwell and Woolf	Timed solenoid valve-injector	18
Fang et al.	More complicated three-layer sandwich injector-commutator with two parallel sample loops, two eluant loops, and two ion-exchange columns on the same rotor	50, 61
Růžička and Hansen	Hydrodynamic injector with part of the carrier stream tubing as a sample loop	58
Basson and van Staden	Carle valve system	19–22
	Valve minder system able of running more than one channel	19, 55
van Staden	Multichannel sample loop injector for simultaneous analysis	Unpublished results

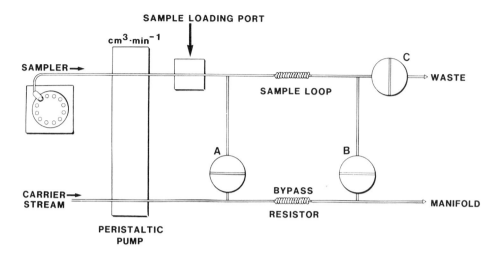

FIGURE 14 Schematic diagram of a sample loop injector with bypass flow resistor.

Following is a brief description of some general techniques used in loop-type injectors.

2.2.1 Sample Loop Injector with Bypass Flow Resistor

This sampling valve system is schematically illustrated in Fig. 14. With valves A and B closed and valve C open, the sample is loaded into the sample loop from the sample source. At the same time the carrier stream bypasses the sample loop and flows through the by-pass resistor to the manifold. To introduce the sample into the carrier stream, the sample loading port and valve C are closed and valves A and B opened. The carrier stream flow in the bypass re-sistor ceases as the hydrodynamic resistance is high enough. At the same moment the carrier stream now flows through the sample loop (in preference to flowing through the resistor) inserting the sample into the analytical manifold. This concept has been used by Růžička and Hansen (10,11,13,14) and nowadays by Tecator (35) in their rotary valve system.

2.2.2 Multiport-Sample-Loop Injectors

This injector type is commercially obtainable in several variants, ranging from low-pressure rotary or slider HPLC-type valves to high-pressure valves. Although not specifically manufactured for FIA, this injector type has been adapted successfully to many FIA systems.

LOAD   INJECT

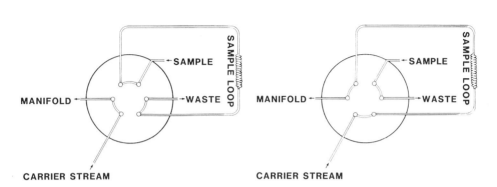

FIGURE 15   Schematic outlay of a one-loop injector.

One-Loop Injector.   The operation of a microvolume one-posi-
tion sampling valve is outlined briefly as follows (Fig. 15).   This
type of valve is usually obtainable as a six-port plug-type valve in
which the sample fills an external loop.   In position 1 (load position)
a sample taken from the sample source is drawn (or pumped) into
the external loop by means of a peristaltic pump.   The carrier
stream flows unobstructed to the manifold system.   Rotation of the
valve rotor to position 2 (inject position) places the sample-filled
loop into the carrier stream flow channel, and the sample is injected
as a plug into the carrier stream.   This single-loop-valve system is
available commercially from Valco, Carle, Rheodyne, and other
manufacturers.

Two-Sample-Loop Injector.   The microvolume two-position sampling valve with two sample loops is obtainable as an 8- or 10-port
plug-type valve (Fig. 16).   While a sample taken from an automated
sampler (or any other sample source) is drawn (or pumped) through
one loop, the carrier stream flows unobstructed to the manifold via
the other loop (loop 2).   When the valve rotor is switched, the sample loop and carrier stream loop alternate and the sample in the
sample-filled loop 1 is now swept by the carrier stream into the
analytical manifold.   At the same time a sample is drawn (or pumped)
through the other loop (loop 2).   This technique is used by Carle,
Valco, and others.

Multichannel-Sample-Loop Injector.   It is also possible to do
simultaneous analysis with FIA, by using valve manipulation with a
multichannel-sample-loop injector.   The multichannel three-layer

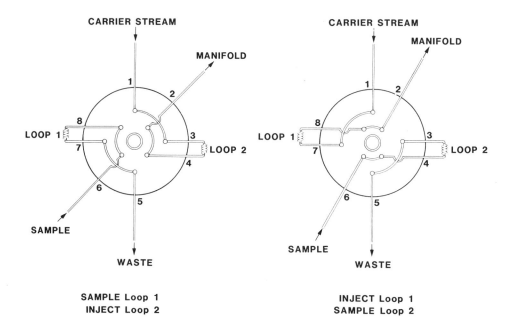

FIGURE 16 Schematic outlay of a double-loop sample injector.

valve designed by the Brazilian research group (39) directed by
Bergamin has been used for the sequential determination of nitrite
and nitrate in the same sample (43) as well as sequential determination of nitrogen and phosphorus in plant material (44). The sliding valve commutator can also be used for simultaneous introduction
of sample and reagent into the same analytical manifold by the merging-zone technique (48), for alternating streams (59), and by zone
trapping (60). A more complicated three-layer sandwich injector-
commutator (50,61) was also utilized for the determination of trace
amounts of heavy metals by the concept of ion-exchange preconcentration. Two parallel sample loops, two eluant loops, and two ion-
exchange columns were arranged on the same rotor. When the rotor
was rotated manually at regular intervals, the sampling, ion-exchange,
and elution processes were done sequentially and alternately on the
two columns.

Jørgensen et al. (51) described a versatile pneumatically-oper-
ated two-layer rotary valve for the diversion of analytical streams
in flow injection analysis. Valve configurations with time-controlled

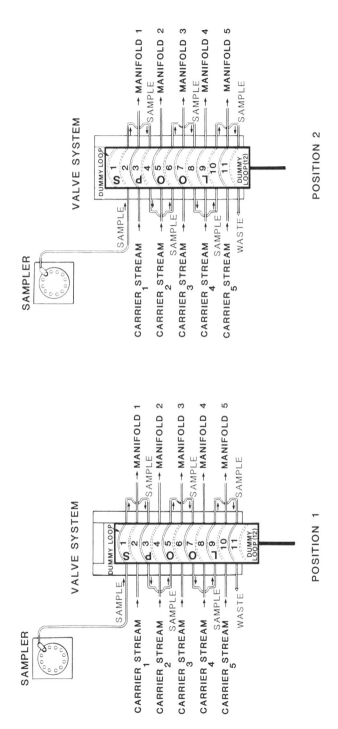

FIGURE 17 Schematic presentation of a multichannel sample loop injector for simultaneous analysis. (From J. F. van Staden, unpublished results.)

sample volumes or with loop-controlled sample volumes in one or two loops and with ion-exchange preconcentration procedures are illustrated. By using a two-loop configuration, sequential sampling is also illustrated (51). The block diagram of a multichannel sample loop injector for simultaneous analysis is outlined in Fig. 17. In position 1, only one sample drawn from the turntable of an automatic sampler is channeled through the dummy loop 1 and loops 3, 5, 7, 9, and 11, while five individual carrier streams analyzing for five different species are channeled via loops 2, 4, 6, 8, and 10 to five different manifold systems. Loop 12 is a dummy. In position 2 the five sample-filled loops mentioned above injected the sample plugs into the five different analytical channels via loops 3, 5, 7, 9, and 11, while loops 2, 4, 6, 8, 10, and the dummy loop 12 are at the same time filled with the next sample. Loop 1 is now a dummy. In position 2 the inside rod is pushed into the outside chamber to align the holes precisely as indicated, while the rod is pulled out in position 1 with the same precautions. If the valve system is properly constructed and carefully adjusted so that the boreholes of the inside rod and outside chamber are precisely aligned in both positions, no pressure buildup is observed. The valve system is rapidly switched by means of compressed air from a gas cylinder without leakage.

2.3   Prevalve and Invalve Sampling Treatment

In the ideal FIA system, samples are presented to the valve system in such a way that the physical and chemical states suit not only the injection part, but also the reaction manifold and detection system. In automated prevalve sampling treatment certain manual sampling preparation procedures can be fully automated or treated to give better results. In the period before the sample is injected into the carrier stream, only the sample is present, so that carryover effects, reagent and sampling mixing dispersion, and so on, are kept to a minimum.

Automated prevalve sample preparation (automated sample pretreatment) may involve physical treatment or chemical modification of samples. This includes

Liquefaction of solid samples
Filtration
Dilution
Digestion
Ion exchange
Preconcentration
Extraction
Dialysis
Chemical treatment

The feasibility of automated prevalve dilution for AAS has been illustrated by Ramsey and Thompson (62) and by Mindel and Karlberg (63). Prevalve automated extraction as an automated sample pretreatment system is suitable for the preconcentration of analyte species when trace components are determined. By using this technique, Nord and Karlberg (64) were able to analyze a number of metal ions with a flow injection atomic absorption detection system. The flow manifold they described allows automatic extraction of metal ions in aqueous samples into 4-methyl-2-pentanone, with ammonium pyrrolidinedithiocarbamate as an extracting agent. The organic extract is led into the loop of an injector situated in an integrated feed system of an atomic absorption spectrometer. This technique was also used by the group of Valcárcel (65). They described a sensitive method for the indirect determination of nitrate and nitrite based on the continuous liquid-liquid extraction of their ion pairs with Cu(I)-neouproine chelate into MIBK, using a FIA-AAS system.

Work has also been reported on invalve sampling treatment in flow injection analysis. Fang et al. (50,61) used the invalve sampling treatment technique on a flow injection system with on-line ion-exchange preconcentration on dual columns for the determination of trace amounts of heavy metals by atomic spectrometry. A multifunctional rotary sampling valve, incorporating two parallel miniature ion-exchange columns packed with a chelating resin with salicylic acid functional groups, was used for sequential sampling, injection, ion exchange, and elution. By using this concept the authors claimed a 20 to 28-fold increase in sensitivity for nickel, copper, lead, and cadmium at a sampling rate of 40 samples per hour compared to direct aspiration of samples.

2.4 Sample Introduction

The efficiency of a FIA system is dependent on the success of its designer in minimizing sample dispersion in all systems, but in such a way that a Gaussian-type peak, if possible, is still obtained. The first contribution to peak variance (peak shape or peak broadening) starts at the injector, where injection volume plays an important role, and from here the dispersion escalates. It is essential that the fluid dynamics before and after sampling remains the same. There are two different cases of injection in flowing streams. In time injection the sample is introduced into the carrier stream for a definite period of time. In an ideal plug injection a slice of the carrier stream is replaced instantaneously by an identical plug of the sample. In both cases the radius, sample length, and initial sample concentration of the injector system contribute to the injector volume and ultimately to the overall peak variance. The theoretical part of time-and-plug injection is discussed in detail in Chapter 2.

In practical FIA systems additional peak broadening may be experienced. This is due to

1. Changes in the cross-sectional area of the flow path in the injector, where changes of fluid velocity and local turbulence will occur.
2. "Dead ends" (dead volume) in the flow path. If a sample reached such a dead end, it will slowly emerge again into the flow stream, causing an e-functional tail in the concentration profile.

In sample loop injection many types of valve designs have been incorporated in flow injection systems. There are many points against which the performance and utility of a particular injector may be assessed for comparative purposes or to judge its usefulness in a particular application. If the FIA system is free of any other interferences and if all the flow streams are stable, high-precision data are determined mainly by a smooth injection system. The required injection volume depends on the type of analytical procedure used and can cover a relatively large range. The amount of sample introduced into an optimized automated procedure is critical and sample loss in the sampling injector should be small. Most injection valve systems have a fixed volume, but if an instrument is to be used for general method development purposes, its injection system should allow continuous volume setting over the wide range needed.

The dead volume of a valve system should be negligible. An ideal valve system should be able for long periods to withstand attack by corrosive reagents used in carrier streams. Corrosive attack from the carrier stream side can be overcome to a certain extent by using water as a carrier stream. It can also be avoided from the sample side by prevalve treatment of corrosive samples. Air bubbles may accumulate to a certain extent in sample loops, before getting loose to be transported into the manifold system. This is very important for high precision and accuracy, because the volume of an air bubble in a loop may make a large contribution to poor results. The formation of air bubbles from carrier stream reagents can be avoided by degassing. Sample injectors should not be labor intensive, but easy to operate to avoid operator faults. This can be achieved by using fully automated sampling systems where individual operator handling is kept to a minimum. The precision and accuracy of automated systems are usually far superior to those of operator-handled systems. Another advantage of microprocessed automated samples is that of the operating conveniences.

3 PUMPING SYSTEMS

To ensure reproducible timing of the movement of the dispersed sample zone through the manifold system, a precise, pulse-free pumping system should be used. The liquid delivery system (pumping system) provides the carrier and sometimes reagent streams at a constant flow rate. A number of techniques in this regard have been described, ranging from the concept of oxidant flow rate (4-6,32), where the driving force is the reduced pressure developed at the nebulizer. Olsen et al. (66) also designed a single-line FIA-AAS manifold where the carrier solution is propelled by a gas-pressurized solvent reservoir. The use of a pump is obviated in both cases. Where an absolute pulse-free flow rate with high precision is possible with a gas-pressurized solvent reservoir, the oxidant flow rate system seems to be less satisfactory. From their experience with high-performance liquid chromatography, Stewart et al. (16,67-71) used a constant-head device, which can be a substrate reservoir (16) with filter, flowmeter, and so on, or recently, a depulsed, positive-displacement pump (Fig. 18) with pulse suppressor operating as a low-to-high-pressure system (17,67-71). Although the positive-displacement pumps are pulse-free with a high precision of delivery, the main disadvantage of these pumps is that more than one pump is required when used in multipump channel analysis.

Most authors prefer to use peristaltic pumps, due to flexibility in number of channels, with a variety of pumping tubes available. Although a rotation of 30 to 40 rpm is preferred (72) by some authors, a 10-rpm pump could also be used, if necessary, with the same precision of final analysis. The lifetime of pumping tubes is normally decreased by pumps rotating at a high speed. Different flexible peristaltic pump tubes with variable inner diameters are available to allow different flow rates and are often color-coded to indicate the flow rate. The pump tubes are made of various materials

1. PVC or Tygon, to be used with aqueous and dilute acid, base, or reagent solution; normally suitable for most FIA applications
2. A PVC derivative (e.g., Solvaflex), which is resistant to most organic solvents but not all
3. Silicone rubber, applicable for some organic materials and acids
4. Fluoroplasts (e.g., Acidflex), which are suitable for concentrated acids and some organic solvents

Peristaltic pumps tend to give slight pulses which become observable with short manifolds (line lengths). This can be eliminated by the introduction of transmission tubes (pulse suppressor coils)

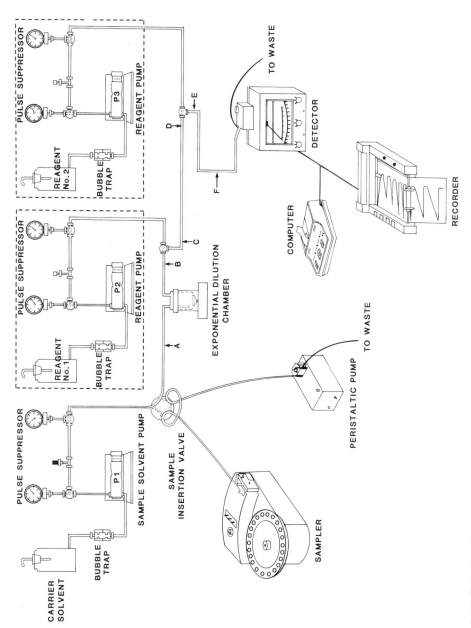

FIGURE 18   Apparatus used for automated multiple-flow injection analysis.   (From Ref. 71.)

between the pump and the valve in the carrier stream and also by placing the tubes between the pump and the analytical manifold in the case of the reagent streams.

4 MANIFOLD SYSTEMS

The reaction manifold system forms the heart of a flow injection analyzer. The manifold system not only determines the accuracy and precision of a FIA system, but indirectly, together with the pumping rate, contributes to the sampling rate of a FIA method. It is not only aspects such as line length and tube diameter that must be taken into consideration when a new procedure is developed, but also aspects such as tube material, connectors, reaction coils, and other manifold components. By careful design of the manifold and judicious choice of flow conditions the dispersion of the injected sample in the analytical conduits can be manipulated to suit the requirements of a large number of analytical procedures.

Reaction manifolds are generally constructed from Tygon, Teflon, or polyethylene tubing with a ±0.5 mm internal diameter the most popular choice, although internal diameters from about 0.2 to 1 mm have been used. With manifold tubing of internal diameter smaller than 0.2 mm there is a tendency toward pressure buildup in the reaction manifold system. This type of manifold is also more easily blocked by particles. If a manifold tubing of more than 1.0 mm is used, the sample throughput normally decreases. If the sample capacity is increased, more reagents are consumed. In the manifold system, tubing of suitable line lengths is wound into coils with varying internal diameter. Incorporation of the correct type of connector into the conduits of reaction manifolds is important to keep the hydrodynamic flow conditions the same before and after the connecting part. The commercially available end fittings and connectors by Tecator (35,72) are suitable for this purpose. The dedicated premanufactured chemical manifolds are called Chemifolds (Fig. 19) by the Tecator group (53).

Various types of flow injection atomic spectroscopy manifold designs are shown schematically in Fig. 20. In the simple single-line manifold [Fig. 20(a)] a single coil of tubing is incorporated between the valve and the nebulizer/expansion chamber (3,6,66, 74-76). It is also possible to use the technique of sample pretreatment, as discussed previously for prevalve and invalve modes, in the single-line manifold system. By using this concept the FIA system is employed for the preparation of suitable solutions prior to their aspiration into the atomic spectroscopy system. Various sample manipulations are possible (63,77). By choosing a suitable volume injected, carrier stream flow rate, tube internal diameter,

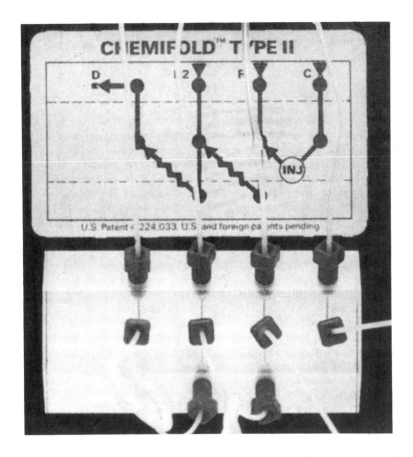

FIGURE 19   Example of a Tecator chemifold. (Reproduced with
the courtesy of Tecator AB, Höganäs, Sweden.)

and tube length, it is possible to achieve appropriate dilution of
samples (3,5,6,9,63,75,78-80) to fall into the limited working range
of atomic absorption spectrometry (75) or other atomic spectroscopic
instrumentation. In this way samples with a very high level of dissolved (63,80) or suspended solids (63) can be processed before
nebulization. Samples with variable viscosity (80) are also analyzed
using this technique. These two types of solutions are difficult to
handle by conventional nebulization (80). Rocks et al. (78) added
lanthanum, which they used as a carrier stream, to samples and
standards with respect to a possible interference effect (81). This
is also achieved by controlled dispersion analysis. Tyson et al.
(81-83) match standards and samples by a standard addition

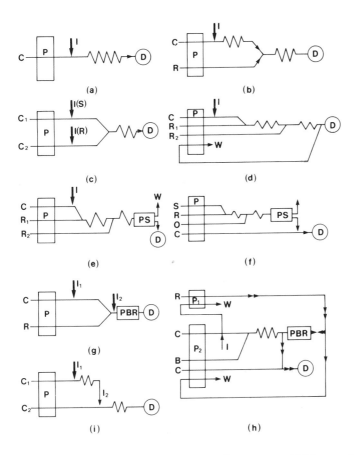

FIGURE 20   Schematic diagrams of some manifolds used in FIA-AAS.
C, Carrier stream; D, detector; I, injection point; O, organic phase;
P, pump; PBR, packed bed reactor; PS, phase separator; R, reagent stream; S, sample solution; W, waste.   Special manifolds:
(a) single line; (b) double line; (c) merging zone; (d) multiline;
(e) and (f) extraction; (g) and (h) ion-exchange; (i) zone sampling.   [Reproduced with permission of the Royal Society of Chemistry (73).]

calibration procedure (81), where samples were used as the carrier
stream and various volumes of the standards were injected (81).
Simultaneous analysis of more than one analyte is also possible by
single-line flow injection analysis.   By using spectrophotometry in
series with atomic absorption spectrometry (Fig. 21), Burguera and
Burguera (84) described a procedure for the simultaneous determin-
ination of iron(II) and total iron.

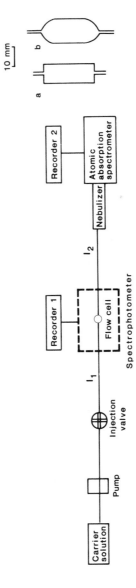

FIGURE 21 Flow diagram of an in-series single-line manifold. The flow-through cells tested (a and b) are shown to the right. [Reproduced with permission of Elsevier Science Publishing (84).]

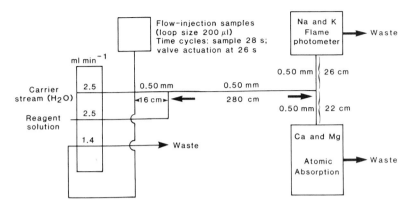

FIGURE 22 Schematic diagram of a double-line manifold split farther downstream into a dual-channel atomic spectrometer and a flame photometer for the simultaneous determination of sodium, potassium, calcium, and magnesium. [Reproduced with permission of Springer-Verlag GmbH & Co. (87).]

Organic solvents are also used as carrier streams in single-line manifold systems (32,85). Taylor and Trevaskis (85) determined lead in gasoline by injecting a treated gasoline sample into a flowing acetone stream for aspiration into an atomic absorption spectrometer.

In a single-line manifold either water or a suitable reagent is used as carrier stream. This can be extended to a double-line manifold [Fig. 20(b)], where a reagent is added farther downstream to the water (or reagent) carrier stream (34,36,63,86). It is also possible to use a multiple-line manifold [Fig. 20(d)], where more reagents are added sequentially to the carrier stream (63). Basson and van Staden (87) split a double-line manifold system farther downstream into a dual-channel atomic spectrometer and a flame photometer for the simultaneous determination of sodium, potassium, calcium, and magnesium (Fig. 22).

It is also possible to introduce sample and reagent solutions separately into carrier streams (e.g., pure water) in such a way that the sampling zone meets the reagent solution zone in a controlled manner (40,46,88). This can be achieved in two different ways: by intermittent pumping (53) or by the use of a multiple injection (39,49,54) with a double proportional sliding valve injector-commutator (37-48) in a merging-zone manifold [Fig. 20(c)]. Reis et al. (48) gave a detailed description of the merging-zone-sampling processes, showing the usefulness of zone sampling in routine work. One of the main advantages of the merging zones concept is that it is extremely useful in saving reagents because

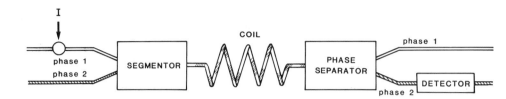

FIGURE 23 Schematic diagram of a continuous flow liquid-liquid extraction system. [Reproduced with permission of Taylor & Francis Ltd. (96).]

the addition (and therefore consumption) of reagent is controlled to merge with the sample zone only.

In the zone-sampling manifold [Fig. 20(i)] a dispersed injected sample zone from valve 1 is sliced by the sample loop of a second valve and introduced into a second carrier stream for transport to the detector (48,73,75). There was an increase in popularity in liquid-liquid extraction [Figs. 20(e) and 20(f)] and ion-exchange manifolds [Figs. 20(g) and 20(h)]. Both manifolds are often the most used in FIA-AAS (or FIA-ICP) for the preconcentration or separation of trace constituents prior to detection.

A number of solvent extraction systems have been described (64,65,80,89-97) and the on-line incorporation of liquid-liquid extraction systems into FIA configurations was also reviewed (96,97). For a complete on-line FIA automatic extraction system, three physical operations are involved

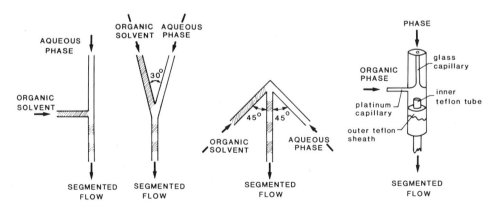

FIGURE 24 Some examples of solvent segmentors. [Reproduced with permission of Taylor & Francis Ltd. (96).]

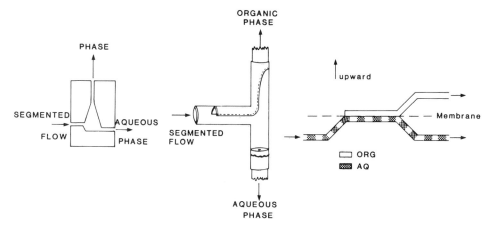

FIGURE 25   Some examples of phase separators.   [Reproduced with permission of Taylor & Francis Ltd. (96).]

1. Phase segmentation
2. Extraction
3. Phase separation

A general arrangement of such an on-line continuous liquid-liquid extraction system is outlined in Fig. 23.   Phase segmentation occurs in a phase segmentor, where the aqueous and organic phases are divided into alternating segments.   A few general types are illustrated in Fig. 24.   The majority of analyte is transferred from one phase to the other (usually from the aqueous phase to the organic phase) and at the same time concentrated in the extraction coil.   Finally, the segmented phases are separated in phase separators (Fig. 25).

By combining a double-line manifold [Fig. 20(b)] with a liquid-liquid extraction manifold [Fig. 20(e)] Burguera and Burguera (89) designed an extraction-flow injection system (Fig. 26) for the rapid determination of cadmium in human urine by extraction with dithizone and FIA-AAS.   They also constructed two-phase separators (Fig. 27) and compared the performance.   In separator a the difference in density of the two phases is used and phase-separating papers employed to complete separation of the phases.   After being forced upward, the aqueous phase passes to waste through a hydrophilic phase-separating paper disk (Whatman, cellulose, 10-mm diameter) held between two washers.   The organic phase is carried into the flow cell after passing through a hydrophobic phase-separating paper disk (silicone-treated Whatman paper, 10-mm diameter) also held between two washers.   In phase separator b the

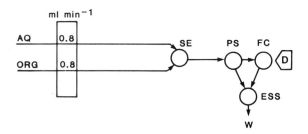

FIGURE 26 Diagram of a double-line extraction-flow injection system. AQ, aqueous phase; ORG, organic phase; SE, segmentor; PS, phase separator; FC, flow cell; D, detector; ESS, exit sucking system; W, waste. [Reproduced with permission of Elsevier Science Publishing (89).]

organic phase is forced downward to the flow cell and the aqueous phase flows to waste. Ogata et al. (90) used a manifold system similar to the previous one (Fig. 26), but used an air compensation method (98) to match the flow rate required by the nebulizer. This is done by spontaneous aspiration of air via a T-connector close to the nebulizer. Nord and Karlberg (64) developed an extraction system which allows automatic extraction prior to the injection valve.

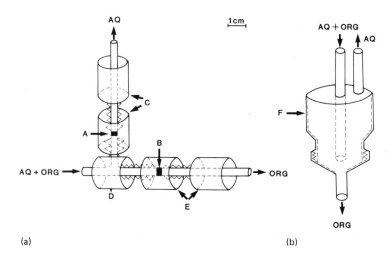

FIGURE 27 Two phase separators, a and b. A, Hydrophilic phase-separating paper; B, hydrophobic phase-separating paper; C, joints for outlet of aqueous phase; D, Daiflon body; E, joints for outlet of organic phase; F, brass body. [Reproduced with permission of Elsevier Science Publishing (89).]

INJECTOR IN THE LOADING POSITION

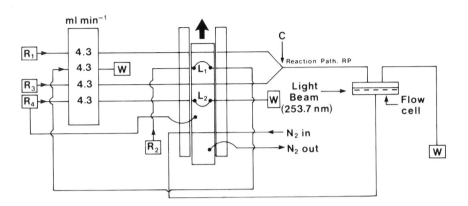

FIGURE 28 Flow injection manifold for the cold vapor atomic absorption technique. L_1, sample loop; L_2, reagent loop; R_1 and R_3, water reservoirs; R_2, sample reservoir; R_4, reagent reservoir; C, confluence point; W, waste. [Reproduced with permission of Pergamon Journals Inc. (88).]

In the procedure the organic extract is led into the loop of an injector situated in an integrated feed system of an atomic absorption spectrometer.

On-line ion-exchange preconcentration [Figs. 20(g) and 20(h)] occurs in two steps (34,66,99–105). In the first step the trace metal ions are preconcentrated on a small ion-exchange column incorporated into a manifold system. Single-, double-, and multiline manifolds are used. The metals are later eluted directly into the nebulizer. Chelex-100 is a popular choice as ion exchanger. Olsen et al. (66) used a single-line two-valve as well as a double-line FIA-AAS system for on-line preconcentration. In the single-line two-valve mode, samples are first injected into the system by valve 1. By closing valve 1 and injecting nitric acid from valve 2, the metal ions are eluted and transported into the AAS instrument. In the double-line FIA-AAS system with directional valve and peristaltic pumps, the metal ions are first preconcentrated and then backflushed into the AAS-instrument. The single-line manifold system is also employed in a FIA-ICP system (99–101).

In the hydride generation technique, analytes such as arsenic, selenium, bismuth, tin, antimony, lead, tellurium, and germanium are converted into volatile species by addition of a solution of sodium borohydride in acidic media. A gas-liquid separator or gas

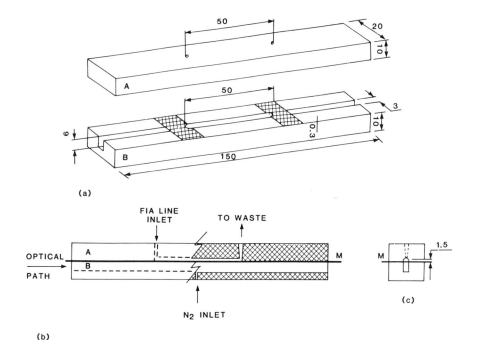

FIGURE 29 Combined phase separator/absorption flow cell: (a) open cell in perspective view; (b) side view with transversal cut; (c) end view. A, top piece; B, bottom piece; M, membrane. All dimensions in millimeters. [Reproduced with permission of Pergamon Journals Inc. (88).]

diffusion cell is used for separation. The gaseous hydride is swept into an atomic absorption spectrometer (106-110,117; G. D. Marshall and J. F. van Staden, unpublished results) or ICP (111) instrument for detection. A manifold similar to the one in Fig. 20(d) is normally used.

It is also possible to determine trace amounts of mercury by combination of flow injection analysis with cold vapor atomic absorption using a Teflon membrane phase separator coupled to the absorption cell (88, 97). The cold vapor atomic absorption technique is based on the chemical reduction of mercury, usually by Sn^{2+} or BH_4^- ions, in a manifold illustrated in Fig. 28. The $Hg°/H_2$ mixture generated during the reduction process is separated from the aqueous carrier solution by a Teflon tape membrane which acts as a gas-liquid phase separator. The commercial Teflon tape is permeable to elementary mercury. The mercury vapor formed in the carrier stream on one side of the Teflon membrane is diffused directly

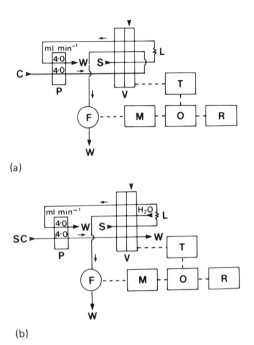

(a)

(b)

FIGURE 30 Flow system and setup for simultaneous multicomponent gradient scanning by flame photometric FIA: (a) calibration curve method; (b) standard addition method. S, sample; C, carrier solution; SC, standard solution carrier; W, waste; L, sample loop; V, valve; P, pump; F, flame nebulizer-burner; T, timer; M, scanning monochromator; O, storage oscilloscope; R, X-Y recorder. Note that the sample loop moves upon injection. [Reproduced with permission of the American Chemical Society (115).]

into the absorption cell positioned in the light path where the absorption of radiation by elementary mercury is measured. The combined phase separator/absorption flow cell is illustrated in Fig. 29.

By combining a merging-zone manifold (see Fig. 18 in Chapter 4) with on-line microwave-oven digestion, Burguera et al. (113) developed a procedure for the determination of copper, zinc, and iron in whole blood. The injected sample is mineralized in the flow system (see Fig. 18 in Chapter 4) on passage through the microwave oven and the metal ions determined by AAS.

The gradient-scanning technique involves rapid scanning of two physical parameters (e.g., wavelength versus absorbance or intensity) at a certain point on the concentration gradient of a well-defined sample zone (114). Fang et al. (115) applied this technique to a flow injection system (Fig. 30) with simultaneous flame

photometric determination of sodium, potassium, lithium, and calcium in tap water and soil extracts. A fast-scanning monochromator and a storage oscilloscope have been employed to obtain spectra
in the range 350 to 800 nm on different sections of the injected sample zone to optimize intensity ranges for all analytes. A novel
standardization method was used to achieve multicomponent standard
addition at different sample/standard ratios in a single injection.

5 NEBULIZER-BURNER SYSTEM

The main purpose of a nebulizer in flame emission, atomic absorption, and inductively coupled plasma is to produce a well-defined
reproducible aerosol for the burner system. However, after nebulization some large droplets remain. Due to incomplete atomization,
a spray chamber is used to condition the aerosol and remove large
droplets. High precision from the nebulizer/spray chamber-burner
system is possible only if the nebulization efficiency can be kept as
high as possible.

5.1 Flame Emission and Atomic Absorption

Only a few applications on flame emission have been reported (40,
87). Various groups, however, used atomic absorption as a detect-
ing device, with the enthusiastic group of Tyson (73,75,80-83,116)
topping the list. It is notable from the flow injection analysis/flame
emission as well as flow injection analysis/atomic absorption literature
that the commercial nebulizer/spray chamber combinations of the commercial instrumentation have been used without modifications (P. L.
Kempster, private discussion). The main efforts to minimize dead
volume were concentrated at the FIA system (73,106).

   The premixed (laminar flow) burners are used mainly in FIA-
AAS systems. Although a number of nebulizers are available, the
concentric pneumatic nebulizer seems to give the best instrumental
performance. The basic mode of action of the concentric pneumatic
nebulizer (116) is shown in Fig. 31. The air at a pressure P_1
(from a suitable compressor) is allowed to escape via a venturi nozzle N. The gas accelerates toward sonic velocity at the venturi
throat, while the pressure drops to a value P_2. As the airstream
broadens beyond the throat, the velocity of air decreases, and at
the same time the air is compressed to a pressure P_3 which is close
to atmospheric pressure. As the pressure P_2 is below atmospheric
pressure, the sample is aspirated into the airstream through the
nebulizer capillary, whose tip is axially positioned in the venturi
tube. The resulting diverging jet of primary aerosol A_1 strikes
the impact bead I and shatters to yield a secondary aerosol A_2,
which progresses through a centrifugal spoiler C (not present in

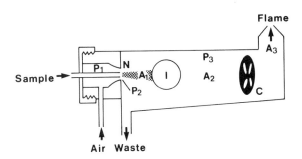

FIGURE 31 Basic mode of action of the concentric pneumatic nebulizer. N, venturi nozzle; I, impact bead; C, centrifugal spoiler; P_1, P_2, and P_3, pressures; A_1, A_2, and A_3, aerosols formed. [Reproduced with permission of the Royal Society of Chemistry (116).]

all commercial designs) before entering the flame as the tertiary aerosol A_3.

A theoretical modeling approach for nebulizer behavior in FIA-AAS is given by Appleton and Tyson (116). The equilibration time (washout time) of the spray chamber in atomic absorption is generally short, but Harnly and Beecher (118) reported that the introduction of samples with high elemental concentrations resulted in contamination of the nebulizer which was not removed during the fill-clean cycle, giving rise to what is called nebulizer memory effects. The sensitivity of an atomic absorption instrument depends on the flow rate of the carrier stream entering the nebulizer and the natural aspiration rate of the nebulizer. Although there is some contradiction as to whether the flow rate should be less (98, 105) or greater (66,119) than the natural aspiration rate of the nebulizer, Tyson (83) shows that although the absorbance increases as the flow rate is increased, the efficiency of the nebulizer (in terms of introducing sample solution into the flame) passes through a maximum (Fig. 32).

5.2 Inductively Coupled Plasma-Atomic Emission Spectrometry

Of the total number of papers devoted to FIA, the amount of papers dealing with FIA-ICP-AES still forms a small fraction (7,66,76,100, 105,111,120-126). Although there has been an increased activity in the utilization of argon-supported ICP-AES as spectroscopic detectors for FIA (120-126), the nebulization techniques used for the introduction of aqueous or organic aerosols into ICP are known to

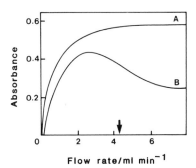

FIGURE 32 Relationship between absorbance and flow rate. (A) Solution continuously pumped into nebulizer; (B) 200 μL injected into carrier stream. [Reproduced with permission of the Royal Society of Chemistry (83).]

limit the powers of achievable detection drastically (120-122). The coupling of FIA to ICP has been accomplished using conventional cross-flow, concentric, or Babington-type pneumatic nebulizers (111, 120,122). The main problem, however, in the achievement of rapid flow analysis when FIA is combined with ICP lies in the slow equilibration times (washout) of the typical conventional ICP spray chambers (127). The large dead volume of spray chambers also contributes to this phenomenon. There are two directions of resolving

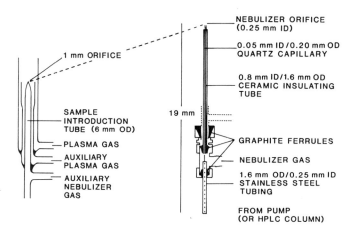

FIGURE 33 Schematic diagram of a microconcentric nebulizer and torch assembly. [Reproduced with permission of the American Chemical Society (122).]

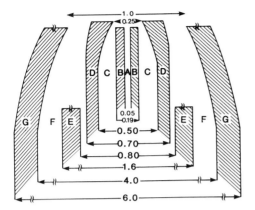

FIGURE 34 Magnified view of the microconcentric nebulizer tip:
A, liquid sample; B, fused-silica inner capillary tube; C, nebulizer
argon gas flow; D, fused-silica outer capillary tube; E, ceramic in-
sulating tube; F, auxiliary nebulizer argon gas; G, normal sample
introduction tube. All dimensions are in millimeters. [Reproduced
with permission of the American Chemical Society (122).]

the slow washout as well as the dead volume problem. The spray
chamber is removed (122) and direct injection nebulization (121,122)
with a microconcentric nebulizer (Fig. 33) is used to introduce the
sample aerosol directly into the plasma. The tip of the nebulizer
gas capillary tube is tapered down to a 0.25-mm orifice as illustrated in Fig. 34. This approach is unfortunately subject to volatiliza-
tion-type interferences, and important analytical figures of merit
remain to be evaluated (121). Another possibility is to reduce the

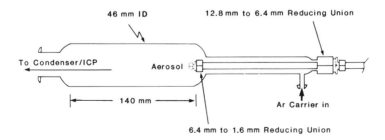

FIGURE 35 Thermospray probe interface. All components were fash-
ioned from standard wall Pyrex tubing except the vaporizer probe
and reducing unions. [Reproduced with permission of the American
Chemical Society (129).]

volume of the chamber sufficiently without removing the spray chamber (112, 128, 130). Koropchak and Winn (129) used a thermospray vaporizer probe (Fig. 35) to interface FIA to the plasma torch. With thermospray the FIA effluent is forced through an electrically heated capillary tube into an expansion chamber. This process provides precisely controlled vaporization of the solution.

6 DATA PROCESSING

The sampling frequency of FIA systems is normally high. Furthermore, a large number of samples are analyzed every day in most routine analytical laboratories. Combining these two factors obviously points to a need for some type of data processing.

6.1 Recorder

Recorders cannot cope with the full data-processing requirements and therefore have only limited usage in modern instrumentation. The visual peaks obtained from strip-chart recorders, however, still form a valuable tool in research work as well as a quick first observation in precision and accuracy of data. Another advantage is that peak form also gives valuable information in research work and sometimes even in routine work.

6.2 Computer

A computer is a flexible instrument to couple with a FIA system. As a source of instant data, the computer serves as a manipulator for peak heights, peak areas, concentration values, and also the statistics thereof, giving a lot of data that may take hours to resolve. The computer is also used in modern instrumentation such as the Tecator system to control various components of such a FIA system.

7. INTERFACING WITH A FIA SYSTEM

It is possible to interface a FIA system in an ideal fashion to an atomic spectroscopic detector, where the detector is capable to

1. Have a fast response to cope with the high sampling rates achieved in FIA systems.
2. Respond linearly over a wide concentration range of components transported to the detector as reaction products from the manifold systems used.

3. Be repeatable over extended periods of time so that frequent calibration runs due to detector weaknesses are not required.
4. Have very low inherent noise characteristics.
5. Have a very low memory effect.
6. Have a relatively high nebulization efficiency.
7. Be sensitive enough to yield the highest possible precision.
8. Be compatible with data processing systems.

7.1 Flame Emission and Atomic Absorption

Flame emission and atomic absorption instrumentation have been successfully interfaced to flow injection. The washout times of spray chambers are short enough to reach equilibrium fast enough to cope with the high sampling rates achieved in FIA. The combination of FIA and flame emission/atomic absorption increases the sampling rate in most cases, and at the same time reduces the amount of sample volume required per analysis. There is also a reduction in the consumption of reagents. Use of the FIA concept as a sample manipulator for flame emission/atomic absorption instrumentation leads to the automation of various manual methods. This includes sample pretreatment procedures such as the addition of releasing agents like lanthanum, as well as more complex manipulations such as liquid-liquid extraction, dilution of samples, processing of samples with a very high level of dissolved or suspended solids, hydride generation, or on-line ion-exchange preconcentration. A prerequisite that is very important in interfacing FIA to flame emission/atomic absorption is the correct combination of carrier stream flow rate and aspiration rate of the nebulizer to yield a maximum sensitivity with a high nebulization efficiency.

7.2 Inductively Coupled Plasma-Atomic Emission Spectrometry

Although FIA is used successfully as a means of sample introduction to an ICP instrument, the unification of these two concepts is not as easy as with AAS. The main reason for this is that the relatively large volume of commercial spray chambers is a source of peak broadening and leads to poor sensitivity and detection limits. The sensitivity can be improved and the memory effect reduced by using spray chambers with reduced volumes. Plasma instability occurs when the spray chamber is removed. As a source ICP is, however, a valuable tool for multicomponent analysis with FIA, as well as performing rapid analyte enrichment/matrix removal.

REFERENCES

1. E. H. Hansen and J. Růžička, The principles of flow injection analysis as demonstrated by three laboratory exercises, *J. Chem. Educ.*, *56*:677 (1979).

2. J. Růžička and E. H. Hansen, Flow injection analysis. Part I. A new concept of fast continuous flow analysis, *Anal. Chim. Acta*, *78*:145 (1975).

3. B. F. Rocks, R. A. Sherwood, and C. Riley, Direct determination of therapeutic concentrations of lithium in serum by flow-injection atomic absorption spectroscopic detection, *Clin. Chem.* *28*:440 (1982).

4. K. Fukamachi and N. Ishibashi, Flow injection atomic absorption spectrometry with organic solvents, *Anal. Chim. Acta*, *119*:383 (1980).

5. J. L. Burguera, M. Burguera, M. Gallignani, and O. M. Alarcón, More on flow injection/atomic absorption analysis for electrolytes, *Clin. Chem.*, *29*:568 (1983).

6. J. L. Burguera, M. Burguera, and M. Gallignani, Direct determination of sodium and potassium in blood serum by flow injection and atomic absorption spectrophotometry, *An. Acad. Bras. Cienc.*, *55*:209 (1983).

7. P. W. Alexander, R. J. Finlayson, L. E. Smythe, and A. Thalib, Rapid flow analysis with inductively-coupled plasma atomic-emission spectroscopy using a micro-injection technique, *Analyst*, *107*:1335 (1982).

8. B. F. Rocks, R. A. Sherwood, and C. Riley, More on flow injection/atomic absorption analysis for electrolytes. Comments, *Clin. Chem.*, *29*:569 (1983).

9. B. F. Rocks, R. A. Sherwood, Z. J. Turner, and C. Riley, Serum iron and total iron-binding capacity determination by flow-injection analysis with atomic absorption detection, *Ann. Clin. Biochem.*, *20*:72 (1983).

10. E. H. Hansen and J. Růžička, Flow injection analysis. Part VI. The determination of phosphate and chloride in blood serum by dialysis and sample dilution, *Anal. Chim. Acta*, *87*:353 (1976).

11. E. H. Hansen, J. Růžička, and B. Rietz, Flow injection analysis. Part VIII. Determination of glucose in blood serum with glucose dehydrogenase, *Anal. Chim. Acta*, *89*:241 (1977).

12. L. Anderson, Simultaneous spectrophotometric determination of nitrite and nitrate by flow injection analysis, *Anal. Chim. Acta*, *110*:123 (1979).

13. J. Růžička, E. H. Hansen, and E. A. G. Zagatto, Flow injection analysis. Part VII. Use of ion-selective electrodes for

rapid analysis of soil extracts and blood serum. Determination of potassium, sodium and nitrate, *Anal. Chim. Acta, 88:*1 (1977).

14. J. Růžička, E. H. Hansen, H. Mosbaek, and F. J. Krug, Pumping pressure and reagent consumption in flow injection analysis, *Anal. Chem., 49:*1858 (1977).

15. S. S. Jørgensen and M. A. B. Regitano, Rapid determination of chromium (VI) by flow injection analysis, *Analyst, 105:*292 (1980).

16. K. K. Stewart, G. R. Beecher, and P. E. Hare, Rapid analysis of discrete samples—use on non-segmented, continuous-flow, *Anal. Biochem., 70:*167 (1976).

17. K. K. Stewart, J. F. Brown, and B. M. Golden, A microprocessor control system for automated multiple flow injection analysis, *Anal. Chim. Acta, 114:*119 (1980).

18. S. D. Rothwell and A. A. Woolf, A timed solenoid injector for flow-analysis, *Talanta, 32:*431 (1985).

19. W. D. Basson and J. F. van Staden, Direct determination of calcium in milk on a non-segmented continuous flow system, *Analyst, 104:*419 (1979).

20. W. D. Basson and J. F. van Staden, Use of non-segmented high-speed continuous flow analysis for the determination of calcium in animal feeds, *Analyst, 103:*296 (1978).

21. W. D. Basson and J. F. van Staden, Low-level determination of hydrazine in boiler feed water with an unsegmented high-speed continuous flow system, *Analyst, 103:*998 (1978).

22. W. D. Basson and J. F. van Staden, Turbidimetric determination of sulfate with a miniaturized manifold on a non-segmented high speed continuous flow system, *Lab. Pract., 27:*863 (1978).

23. B. Karlberg and S. Thelander, Determination of readily oxidised compounds by flow injection analysis and redox potential detection, *Analyst, 103:*1154 (1978).

24. A. H. T. M. Scholten, U. A. Th. Brinkman, and R. W. Frei, Photochemical reaction detectors in continuous-flow systems. Applications to pharmaceuticals, *Anal. Chim. Acta, 114:*137 (1980).

25. V. V. S. E. Dutt, A. E. Hanna, and H. A. Mottola, Kinetically assisted equilibrium-based repetitive determination of iron(II) with ferrozine in flow-through systems, *Anal. Chem., 48:*1207 (1976).

26. V. V. S. E. Dutt, D. Scheeler, and H. A. Mottola, Repetitive determinations of iron(III) in closed flow-through systems by series reactions, *Anal. Chim. Acta, 94:*289 (1977).

27. R. C. Schothorst, J. M. Reijn, H. Poppe, and G. den Boef, The application of strongly reducing agents in flow injection analysis. Part 1. Chromium(II) and vanadium(II), *Anal. Chim. Acta*, *145*:197 (1983).

28. R. C. Schothorst and G. den Boef, The application of strongly reducing agents in flow injection analysis. Part 2. Chromium(II), *Anal. Chim. Acta*, *153*:133 (1983).

29. R. C. Schothorst, J. J. F. van Veen, and G. den Boef, The application of strongly reducing agents in flow injection analysis. Part 3. Vanadium(II), *Anal. Chim. Acta*, *161*:27 (1984).

30. R. C. Schothorst, M. van Son, and G. den Boef, The application of strongly reducing agents in flow injection analysis. Part 4. Uranium(III), *Anal. Chim. Acta*, *162*:1 (1984).

31. J. W. Keller, Enzyme assay by repetitive flow injection analysis. Application to the assay of dog kidney aminoacylase, *Anal. Lett.*, *17*:589 (1984).

32. A. S. Attiyat and G. D. Christian, Nonaqueous solvents as carrier or sample solvent in flow analysis/atomic absorption spectrometry, *Anal. Chem.*, *56*:439 (1984).

33. M. J. Medina, J. Bartroli, J. Alonso, M. Blanco, and J. Fuentes, Direct determination of glucose in blood serum using Trinder's reaction, *Anal. Lett.*, *17*:385 (1984).

34. F. Malamas, M. Bengtsson, and G. Johansson, On-line trace metal enrichment and spectrometry by a column containing immobilized 8-quinolinol in a flow-injection system, *Anal. Chim. Acta*, *160*:1 (1984).

35. B. I. Karlberg, Automation of wet chemical procedures using FIA, *Am. Lab.*, *15*:73 (1983).

36. P. Martínez-Jiménez, M. Gallego, and M. Valcárcel, Indirect atomic absorption determination of uranium and lanthanum by flow injection analysis using an air-acetylene flame, *At. Spectrosc.*, *6*:137 (1985).

37. H. Bergamin F[O]., J. X. Medeiros, B. F. Reis, and E. A. G. Zagatto, Solvent extraction in continuous flow analysis. Determination of molybdenum in plant material, *Anal. Chim. Acta*, *101*:9 (1978).

38. H. Bergamin F[O]., B. F. Reis, A. O. Jacintho, and E. A. G. Zagatto, Ion exchange in flow injection analysis. Determination of ammonium ions at the $\mu g \ L^{-1}$ level in natural waters with pulsed Nessler reagent, *Anal. Chim. Acta*, *117*:81 (1980).

39. H. Bergamin F[O]., E. A. G. Zagatto, F. J. Krug, and B. F. Reis, Merging zones in flow injection analysis. Part 1. Double proportional injectors and reagent consumption, *Anal. Chim. Acta*, *101*:17 (1978).

40. E. A. G. Zagatto, F. J. Krug, H. Bergamin F⁰., S. S. Jørgensen, and B. F. Reis, Merging zones in flow injection analysis. Part 2. Determination of calcium, magnesium and potassium in plant material by continuous flow injection atomic absorption and flame emission spectrometry. *Anal. Chim. Acta* 104:279 (1979).

41. B. F. Reis, H. Bergamin F⁰., E. A. G. Zagatto, and F. J. Krug, Merging zones in flow injection analysis. Part 3. Spectrophotometric determination of aluminium in plant and soil materials with sequential addition of pulsed reagents, *Anal. Chim. Acta, 107*:309 (1979).

42. E. A. G. Zagatto, B. F. Reis, H. Bergamin F⁰., and F. J. Krug, Isothermal distillation in flow injection analysis, Determination of total nitrogen in plant material, *Anal. Chim. Acta, 109*:45 (1979).

43. M. F. Giné, H. Bergamin F⁰., E. A. G. Zagatto, and B. F. Reis, Simultaneous determination of nitrate and nitrite by flow injection analysis, *Anal. Chim. Acta, 114*:191 (1980).

44. B. F. Reis, E. A. G. Zagatto, A. O. Jacintho, F. J. Krug, and H. Bergamin F⁰., Merging zones in flow injection analysis. Part 4. Simultaneous spectrophotometric determination of total nitrogen and phosphorus in plant material, *Anal. Chim. Acta, 119*:305 (1980).

45. E. A. G. Zagatto, A. O. Jacintho, J. Mortalti, and H. Bergamin F⁰., An improved flow injection determination of nitrite in waters by using intermittent flows, *Anal. Chim. Acta, 120*:399 (1980).

46. E. A. G. Zagatto, A. O. Jacintho, L. C. R. Pessenda, F. J. Krug, B. F. Reis, and H. Bergamin F⁰., Merging zones in flow injection analysis. Part 5. Simultaneous determination of aluminium and iron in plant digest by a zone-sampling approach, *Anal. Chim. Acta, 125*:37 (1981).

47. F. J. Krug, J. Mortalti, L. C. R. Pessenda, E. A. G. Zagatto, and H. Bergamin F⁰., Flow injection spectrophotometric determination of boron in plant material with azomethine-H, *Anal. Chim. Acta, 125*: 29 (1981).

48. B. F. Reis, A. O. Jacintho, J. Mortalti, F. J. Krug, E. A. G. Zagatto, H. Bergamin F⁰., and L. C. R. Pessenda, Zone-sampling processes in flow injection analysis, *Anal. Chim. Acta, 123*:221 (1981).

49. J. Mindegaard, Flow multi-injection analysis. A system for the analysis of highly concentrated samples without prior dilution, *Anal. Chim. Acta, 104*:185 (1979).

50. Z. Fang, J. Ruzicka, and E. H. Hansen, An efficient flow-injection system with on-line ion-exchange preconcentration for the determination of trace amounts of heavy metals by atomic absorption spectrometry, *Anal. Chim. Acta, 164*:23 (1984).

51. S. S. Jørgensen, K. M. Petersen, and L. A. Hansen, A simple multifunctional valve for flow injection analysis, *Anal. Chim. Acta*, *169*:51 (1985).

52. H. Baadenhuijsen and H. E. H. Seuren-Jacobs, Determination of total CO_2 in plasma by automated flow injection analysis, *Clin. Chem.*, *25*:443 (1979).

53. J. Růžička and E. H. Hansen, Flow injection analysis. Principles, applications and trends, *Anal. Chim. Acta*, *114*:19 (1980).

54. J. Růžička and E. H. Hansen, Stopped flow and merging zones. A new approach to enzymatic assay by flow injection analysis, *Anal. Chim. Acta*, *106*: 207 (1979).

55. J. F. van Staden, Simultaneous determination of protein (nitrogen), phosphorus, and calcium in animal feeds by multichannel flow injection analysis, *J. Assoc. Off. Anal. Chem.*, *106*:718 (1983).

56. C. Riley, L. H. Aslett, B. F. Rocks, R. A. Sherwood, J. D. McK. Watson, and J. Morgon, Controlled dispersion analysis: Flow-injection analysis without injection, *Clin. Chem.*, *29*:332 (1983).

57. R. A. Sherwood, B. F. Rocks, and C. Riley, Controlled-dispersion flow-analysis with atomic-absorption detection for the determination of clinically relevant elements, *Analyst*, *110*:493 (1985).

58. J. Růžička and E. H. Hansen, Recent developments in flow injection analysis: Gradient techniques and hydrodynamic injection, *Anal. Chim. Acta*, *145*:1 (1983).

59. H. Kagenow and A. Jensen, Differential kinetic analysis and flow injection analysis. Part 3. The (2.2.2) cryptates of magnesium, calcium and strontium, *Anal. Chim. Acta*, *114*:227 (1980).

60. A. Tanaka, M. Miyazaki, and T. Deguchi, New simultaneous catalytic determination of thiocyanate and iodide by flow injection analysis, *Anal. Lett.*, *18*:695 (1985).

61. Z. Fang, S. Xu, and S. Zhang, The determination of trace amounts of heavy metals in waters by a flow-injection system including ion-exchange preconcentration and flame atomic absorption spectrometric detection, *Anal. Chim. Acta*, *164*:41 (1984).

62. M. H. Ramsey and M. Thompson, Online diluter for atomic-absorption spectrophotometry, *Analyst*, *107*:232 (1982).

63. B. D. Mindel and B. Karlberg, A sample pretreatment system for atomic absorption using flow injection analysis, *Lab. Pract.*, *30*:719 (1981).

64. L. Nord and B. Karlberg, Sample preconcentration by continuous flow extraction with a flow injection atomic absorption detection system, *Anal. Chim. Acta*, *145*:151 (1983).

65. M. Gallego, M. Silva, and M. Valcárcel, Determinación of nitrate and nitrite by continuous liquid-liquid extraction with a flow injection atomic absorption detection system, *Fresenius Z. Anal. Chem.*, *323:*50 (1986).

66. S. Olsen, L. C. R. Pessenda, J. Růžička, and E. H. Hansen, Combination of flow injection analysis with flame atomic-absorption spectrophotometry: Determination of trace amounts of heavy metals in polluted seawater, *Analyst*, *108:*905 (1983).

67. K. K. Stewart, Depulsing system for positive displacement pumps, *Anal. Chem.*, *49:*2125 (1977).

68. K. K. Stewart and A. G. Rosenfeld, Automated titrations: The use of automated multiple flow injection analysis for the titration of discrete samples, *J. Autom. Chem.*, *3:*30 (1981).

69. J. F. Brown, K. K. Stewart, and D. Higgs, Microcomputer control and data system for automated multiple flow injection analysis, *J. Autom. Chem.*, *3:*182 (1981).

70. K. K. Stewart, Flow injection analysis. New tool for old assays. New approach to analytical measurements, *Anal. Chem.*, *55:*931A (1983).

71. K. K. Stewart and A. G. Rosenfeld, Exponential dilution chambers for scale expansion in flow injection analysis, *Anal. Chem.*, *54:*2368 (1982).

72. J. Růžička and E. H. Hansen, *Flow Injection Analysis*, Wiley, New York, 1981.

73. J. F. Tyson, Flow-injection analysis techniques for atomic-absorption spectrometry—a review, *Analyst*, *110:*419 (1985).

74. W. R. Wolf and K. K. Stewart, Automated multiple flow injection analysis for flame atomic absorption spectrometry, *Anal. Chem.*, *51:*1201 (1979).

75. J. F. Tyson, Flow injection techniques for flame atomic absorption spectrophotometry, *Trends Anal. Chem.*, *4:*124 (1985).

76. C. W. McLeod, P. J. Worsfold, and A. G. Cox, Simultaneous multi-element analysis of blood serum by flow injection-inductively coupled plasma atomic emission spectrometry, *Analyst*, *109:*327 (1984).

77. N. Zhou, W. Frech, and E. Lundberg, Rapid determination of lead, bismuth, antimony and silver in steels by flame atomic absorption spectrometry combined with flow injection analysis, *Anal. Chim. Acta*, *153:*23 (1983).

78. B. F. Rocks, R. A. Sherwood, and C. Riley, Direct determination of calcium and magnesium in serum using flow injection analysis and atomic absorption spectroscopy, *Ann. Clin. Biochem.*, *21:*51 (1984).

79. B. F. Rocks, R. A. Sherwood, L. M. Bayford, and C. Riley, Zinc and copper determination in microsamples of serum by flow injection and atomic absorption spectroscopy, Ann. Clin. Biochem., 19:338 (1982).

80. J. F. Tyson, C. E. Adeeyinwo, J. M. H. Appleton, S. R. Bysouth, A. B. Idris, and L. L. Sarkissian, Flow injection techniques of method development for flame atomic absorption spectrometry, Analyst, 110:487 (1985).

81. J. F. Tyson, J. M. H. Appleton, and A. B. Idris, Flow injection calibration methods for atomic absorption spectrometry, Anal. Chim. Acta, 145:159 (1983).

82. J. F. Tyson and A. B. Idris, Flow injection sample introduction for atomic-absorption spectrometry: Applications of a simplified model for dispersion, Analyst, 106:1125 (1981).

83. J. F. Tyson, Low cost continuous flow analysis. Flow-injection techniques in atomic-absorption spectrometry, Anal. Proc. (London), 18:542 (1981).

84. J. L. Burguera and M. Burguera, Flow injection spectrophotometry followed by atomic absorption spectrometry for the determination of iron(II) and total iron, Anal. Chim. Acta, 161:375 (1984).

85. C. G. Taylor and J. M. Trevaskis, Determination of lead in gasoline by a flow-injection technique with atomic absorption spectrometric detection, Anal. Chim. Acta, 179:491 (1986).

86. B. W. Renoe, C. E. Shideler, and J. Savory, Use of a flow-injection sample manipulator as an interface between a "high-performance" liquid chromatograph and an atomic absorption spectrometer, Clin. Chem., 27:1546 (1981).

87. W. D. Basson and J. F. van Staden, Simultaneous determination of sodium, potassium, magnesium and calcium in surface, ground and domestic water by flow injection analysis, Fresenius Z. Anal. Chem., 302:370 (1980).

88. J. C. de Andrade, C. Pasquini, N. Ballan, and J. C. van Loon, Cold vapor atomic absorption determination of mercury by flow injection analysis using a Teflon membrane phase separator coupled to the absorption cell, Spectrochim. Acta, Part B, 38:1329 (1983).

89. J. L. Burguera and M. Burguera, Determination of cadmium in human urine by extraction with dithizone in a flow injection system, Anal. Chim. Acta, 153:207 (1983).

90. K. Ogata, S. Tanabe, and T. Imanari, Flame atomic absorption spectrophotometry coupled with solvent extraction/flow injection analysis, Chem. Pharm. Bull., 31:1419 (1983).

91. M. Gallego and M. Valcárcel, Indirect atomic absorption spectrometric determination of perchlorate by liquid-liquid extraction in a flow-injection system, Anal. Chim. Acta, 169:161 (1985).

92. J. A. Sweleh and F. F. Cantwell, Sample introduction by solvent extraction/flow injection to eliminate interferences in atomic absorption spectroscopy, *Anal. Chem.*, *57*:420 (1985).

93. L. Nord and B. Karlberg, An automated extraction system for flame atomic absorption spectrometry, *Anal. Chim. Acta*, *125*:199 (1981).

94. K. Bäckström, L. G. Danielsson, and L. Nord, Sample work-up for graphite-furnace atomic-absorption spectrometry using continuous-flow extraction, *Analyst*, *109*:323 (1984).

95. M. Bengtsson and G. Johansson, Preconcentration and matrix isolation of heavy metals through a two-stage solvent extraction in a flow system, *Anal. Chim. Acta*, *158*:147 (1984).

96. M. D. Luque de Castro, Flow injection analysis: A new tool to automate extraction processes, *J. Autom. Chem.*, *8*:56 (1986).

97. Z. Fang, S. Xu, X. Wang, and S. Zhang, Combination of flow-injection techniques with atomic spectrometry in agricultural and environmental analysis, *Anal. Chim. Acta*, *179*:325 (1986).

98. N. Yoza, Y. Aoyagi, and S. Ohashi, Flow injection system for atomic absorption spectrometry, *Anal. Chim. Acta*, *111*: 163 (1979).

99. I. G. Cook, C. W. Mcleod, and P. J. Worsfold, Use of activated alumina as a column packing material for adsorption of oxyanions in flow injection analysis with ICP-AES (inductively coupled plasma-atomic emission spectrometry) detection, *Anal. Proc.*, *23*:5 (1986).

100. C. W. Mcleod, I. G. Cook, P. J. Worsfold, J. E. Davies, and J. Queay, Analyte enrichment and matrix removal in flow-injection analysis-inductively coupled plasma-atomic emission-spectrometry. Determination of phosphorus in steels, *Spectrochim. Acta Part B*, *40*:57 (1985).

101. A. T. Haj-Hussein, G. D. Christian, and J. Růžička, Determination of cyanide by atomic absorption using a flow injection conversion method, *Anal. Chem.*, *58*:38 (1986).

102. A. G. Cox and C. W. Mcleod, Preconcentration and determination of trace chromium(III) by flow injection/inductively-coupled plasma/atomic emission spectrometry, *Anal. Chim. Acta*, *179*:487 (1986).

103. E. B. Milosavljevic, J. Růžička, and E. H. Hansen, Simultaneous determination of free and EDTA-complexed copper ions by flame atomic absorption spectrometry with an ion-exchange flow-injection, *Anal. Chim. Acta*, *169*:321 (1985).

104. M. A. Marshall and H. A. Mottola, Performance studies under flow conditions of silica-immobilized 8-quinolinol and its

application as a preconcentration tool in flow injection/atomic absorption determinations, *Anal. Chem.*, *57*:729 (1985).

105. O. F. Kamson and A. Townshend, Ion-exchange removal of some interferences on the determination of calcium by flow injection analysis and atomic absorption spectrometry, *Anal. Chim. Acta*, *155*:253 (1983).

106. O. Åström, Flow injection analysis for the determination of bismuth by atomic absorption spectrometry with hydride generation, *Anal. Chem.*, *54*:190 (1982).

107. G. E. Pacey, M. R. Straka, and J. R. Gord, Dual phase gas diffusion flow injection analysis/hydride generation atomic absorption spectrometry, *Anal. Chem.*, *58*:502 (1986).

108. H. Narasaki and M. Ikeda, Automated determination of arsenis and selenium by atomic absorption spectrometry with hydride generation, *Anal. Chem.*, *57*:1382 (1985).

109. M. Yamamoto, M. Yasuda, and Y. Yamamoto, Hydride-generation atomic absorption spectrometry coupled with flow injection analysis, *Anal. Chem.*, *57*:1382 (1985).

110. X. Wang and Z. Fang, The determination of arsenic in environmental samples by flow injection hydride generation atomic absorption spectrometry, *Fenxi Huaxue*, in press.

111. S. Greenfield, Inductively coupled plasma-atomic emission spectroscopy (ICP-AES) with flow injection analysis (FIA), *Spectrochim. Acta Part B*, *38*:93 (1983).

112. P. L. Kempster, J. F. van Staden, and H. R. van Vliet, Investigation of small volume cloud chambers for use in inductively coupled plasma nebulization, *J. Anal. At. Spectrom.* *2*:823 (1987).

113. M. Burguera, J. L. Burguera, and O. M. Alarcón, Flow injection and microwave-oven sample decomposition for determination of copper, zinc and iron in whole blood by atomic absorption spectrometry, *Anal. Chim. Acta*, *179*:351 (1986).

114. J. Růžička and E. H. Hansen, Recent developments in flow injection analysis: Gradient techniques and hydrodynamic injection, *Anal. Chim. Acta*, *145*:1 (1983).

115. Z. Fang, J. M. Harris, J. Růžička, and E. H. Hansen, Simultaneous flame photometric determination of lithium, sodium, potassium, and calcium by flow injection analysis with gradient scanning standard addition, *Anal. Chem.*, *57*:1457 (1985).

116. J. M. H. Appleton and J. F. Tyson, Flow injection atomic absorption spectrometry: The kinetics of instrument response, *J. Anal. At. Spectrom.*, *1*:63 (1986).

117. C. C. Y. Chan, Semiautomated method for determination of selenium in geological materials using a flow injection analysis technique, *Anal. Chem.*, *57*:1482 (1985).

118. J. M. Harnly and G. R. Beecher, Two-valve injector to minimize nebulizer memory for flow injection atomic absorption spectrometry, *Anal. Chem.*, *57*:2015 (1985).

119. M. W. Brown and J. Růžička, Parameters affecting sensitivity and precision in the combination of flow injection analysis with flame atomic-absorption spectrophotometry, *Analyst*, *109*: 1091 (1984).

120. S. Greenfield, FIA (flow injection analysis) weds ICP (inductively coupled plasma)—a marriage of convenience, *Ind. Res. Dev.*, *23*:140 (1981).

121. K. E. LaFremiere, G. W. Rice, and V. A. Fassel, Flow injection analysis with inductively coupled plasma-atomic emission spectroscopy. Critical comparison of conventional pneumatic, ultrasonic and direct injection nebulization, *Spectrochim. Acta Part B*, *40*:1495 (1985).

122. K. E. Lawrence, G. W. Rice, and V. A. Fassel, Direct liquid sample introduction for flow injection analysis and liquid chromatography with inductively coupled argon plasma spectrometric detection, *Anal. Chem.*, *56*:289 (1984).

123. Y. Israel and R. M. Barnes, Standard addition method in flow injection analysis with inductively coupled plasma atomic emission spectrometry, *Anal. Chem.*, *56*:1188 (1984).

124. R. R. Liversage, J. C. van Loon, and J. C. Andrede, A flow injection/hydride generation system for the determination of arsenic by inductively-coupled plasma atomic emission spectrometry, *Anal. Chim. Acta*, *161*:275 (1984).

125. J. A. C. Broekaert and F. Leis, An injection method for the sequential determination of boron and several metals in wastewater samples by inductively-coupled plasma atomic emission spectrometry, *Anal. Chim. Acta*, *109*:73 (1979).

126. E. A. G. Zagatto, A. O. Jacintho, F. J. Krug, B. F. Reis, R. E. Bruns, and M. C. U. Araújo, Flow injection systems with inductively-coupled argon plasma atomic emission spectrometry. Part 2. The generalized standard addition method, *Anal. Chim. Acta*, *145*:169 (1983).

127. R. F. Browner and A. W. Boorn, Sample introduction techniques for atomic spectroscopy, *Anal. Chem.*, *56*:875A (1984).

128. T. Ito, E. Nakagawa, H. Kawaguchi, and A. Mizuike, Semiautomatic microliter sample injection into an inductively coupled plasma for simultaneous multielement analysis, *Mikrochim. Acta*, *1*: 423 (1982).

129. J. A. Koropchak and D. H. Winn, Thermospray interfacing for flow injection analysis with inductively coupled plasma atomic emission spectrometry, *Anal. Chem.*, *58:*2558 (1986).

130. P. L. Kempster, J. F. van Staden, and H. R. van Vliet, Determination of calcium in water by flow injection analysis-inductively coupled plasma (FIA-ICP) emission spectrometry, *Fresenius Z. Anal. Chem. 332:*153 (1988).

4

Analytical Methods and Techniques

ZHAOLUN FANG *Flow Injection Analysis Research Center, Institute of Applied Ecology, Academia Sinica, Shenyang, Liaoning, China*

1 INTRODUCTION

The combination of flow injection analysis with atomic spectroscopy
has created an entire array of new analytical methods and techniques which have considerably extended the capabilities of the
conventional atomic spectrometric methods (particularly flame atomic
absorption and induction coupled plasma spectrometric methods) in
terms of efficiency, sensitivity, saving of sample and reagents,
freedom from interference, and so on. To date, the function of
these improvements belong mostly to the following three categories:

1. Microsample introduction
2. On-line sample pretreatment (dilution, separation, reagent addition, digestion, etc.)
3. Novel calibration methods

However, recent developments show that the potentialities
are far from being fully explored and it seems reasonable to expect
that the high versatility of the flow injection technique will continue
to stimulate interest in the creation of more and better methods and
techniques. In this chapter the author provides the reader with
an overview of the currently available methods and techniques in
flow injection atomic spectroscopy. Methods on separation are not
treated in detail in this chapter, as they are dealt with in Chapter 5.

2 SAMPLE INTRODUCTION TECHNIQUES

Because of its simplicity and reliability, liquid sample introduction
is undoubtedly the most widely used form of sample introduction in
atomic spectroscopy, yet it has been and still is considered as the
weakest link or the "Achilles' heel" of atomic spectroscopy (1). Low
transport efficiency, relatively high sample consumption, and low
tolerance of solute concentration are some of the most important aspects of conventional sample introduction which require improvement. The advent of flow injection analysis as a new concept in
solution handling has provided atomic spectroscopy with a new form
of sample introduction which could partially overcome some of the
shortcomings in conventional sample introduction. General features
of flow injection sample introduction include

1. Considerable decrease in sample consumption, normally using
   10 to 200 µL of sample, in contrast to a few milliliters in the
   conventional mode with little loss in sensitivity; this feature
   is especially important when analyzing precious samples such
   as blood serum.

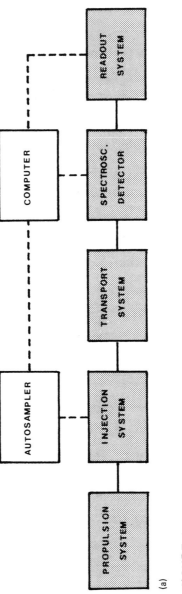

(a)

FIGURE 1

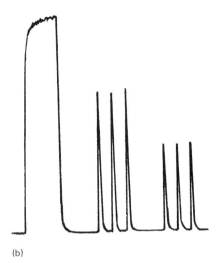

(b)

FIGURE 1 (a) Schematic diagram of flow injection sample introduction manifold for atomic spectroscopic analysis. (b) Typical
recordings of flow injection sample introduction with atomic spectroscopic detection. From right to left: triplicate injections of
50-μL sample volume and 100-μL sample volume using a short mixing coil as transport system compared to conventional sample
introduction.

2. High tolerance of salt content in samples (see Section 2.5).
3. Depression of matrix effects due to differences in viscosity of
   sample and standard (see Section 2.6).
4. High sample throughput of 100 to 700 samples per hour (see
   Section 2.4).

2.1 Manifold Configurations for Flow Injection Sample Introduction

The basic configuration for a flow injection sample introduction manifold is really very simple, consisting of a sample propulsion system,
an injection system, a transport system, the spectrometric detector,
and the readout system [Fig. 1(a)]. A defined volume (10 to 200
μL) of sample is inserted reproducibly as a plug into a flowing carrier stream via the injection system (usually, an injection valve, or
less often, a septum injection port) and transported through the
transport system into the nebulizer of the spectrometer, where the
sample and carrier streams are nebulized and atomized continuously;
the resulting absorption or emission signal output in the form of a
transient peak is then recorded by a chart recorder or processed

by a microcomputer. Carrier solutions are usually propelled by
peristaltic pumps, but high-performance liquid chromatographic
(HPLC) solvent delivery systems have also been used (2-4). In
the simplest form, the carrier could be propelled by the suction
action of the nebulizer, although at the expense of losing some of
the favorable features of flow injection sample introduction, including independent control of carrier flow rate and the minimizing of
effects due to viscosity differences. The transport system might
also be as simple as a 10-cm length of 0.5- to 1-mm-ID Tygon or
Teflon tubing connecting the injection valve and the nebulizer.
With the necessary components at hand, a simple sample introduction manifold may be set up in a very short time and connected to
the spectrometer or disconnected from it in a few minutes. In more
sophisticated manifolds where a high level of automation is desired,
an automatic injection valve may be used which could control the
sampling and injection periods. An autosampler might also be included and its action synchronized with the injection valve via a
timer or microprocessor. Typical recordings of a flow injection
sample introduction atomic absorption system are shown in Fig. 1(b)
compared to signals obtained with conventional sample introduction.
The most outstanding difference is the high sampling frequency obtainable in the flow injection mode; however, one will of course
also notice the loss in sensitivity.

2.2 Sensitivity of Flow Injection Sample Introduction

One of the criticisms often leveled at flow injection sample introduction in atomic spectroscopy is its low relative sensitivity expressed
in characteristic concentration. Admittedly, the recorded transient
peak height most often used for evaluation using the flow injection
mode of sample introduction is only a fraction of the signal obtained
using conventional aspiration and measuring the steady-state signal.
This unfavorable feature could not possibly be avoided if the important merits of the flow injection mode of operation, including
microliter sample volumes and high sampling frequency, were to be
retained. Yet with a short connection tube of about 20 cm for the
transport system, approximately 60% of the steady-state signal could
be attained with 100 μL of sample at a reasonably high sampling frequency of 150 samples per hour and 15% with 20 μL of sample, allowing high sampling rates of 200 to 500 samples per hour. Therefore,
unless the highest sensitivity is desired, in cases where the determination is made close to the detection limit, it is unlikely that the
loss in sensitivity will seriously degrade the quality of the analytical result. On the other hand, even higher relative sensitivity
could be achieved by using more concentrated samples which could
not be analyzed without dilution in the conventional mode of sample

introduction (see Section 2.4). The large tolerance of salt content in samples using flow injection sample introduction is a factor that should be considered during the sample preparation stage so that the loss in sensitivity could be well compensated for by decreasing the dilution factor whenever possible. Thus Zhou et al. (5) have improved the detection limit for antimony, arsenic, bismuth, and lead in steel samples by using an eightfold more concentrated sample solution in a flow injection atomic absorption system. Fang et al. (6) have reported a threefold gain in sensitivity in the analysis of copper in fertilizers by using highly concentrated sample solutions.

Whereas special measures as that quoted above are necessary to attain a relative sensitivity similar to conventional sample introduction, the absolute sensitivity is almost always four- to eightfold higher with flow injection sample introduction (20 to 100 μL injection volume). This is quite natural, as the sample consumption for flow injection is much smaller.

Wolf and Stewart (3) and Harnly and Beecher (7) have shown that when peak area was used as the readout mode, the carrier flow rate of a flow injection sample introduction system with atomic absorption detection had considerable influence on the sensitivity due to variations in the nebulization efficiency. Higher efficiency could be obtained at lower flow rates. Thus when 25 to 300 μL of a sample of 2 μg of zinc per milliliter was injected, the sensitivity was improved more than onefold when the carrier flow rate was decreased from 10 to 5 mL min^{-1} (7). This improvement could of course only be attained at a sacrifice in the sampling frequency because of the lower pumping rate.

The sensitivity of flow injection sample introduction with atomic absorption detection could also be enhanced by using an appropriate organic solvent as carrier. Fukamachi and Ishibashi (8) were able to almost double the sensitivity in the determination of copper in an aqueous sample by using *n*-butyl acetate or methyl isobutyl ketone (MIBK) as carrier. Attiyat and Christian (9), after comparing different combinations of organic solvent carriers and sample solvents, have shown that an optimum combination of MIBK as carrier and acetone as sample solvent could bring about an eightfold enhancement in sensitivity compared to aqueous systems.

2.3 Precision and Detection Limits for Flow Injection Sample Introduction

Contradictory evaluations on the precision and relative detection limits of flow injection sample introduction as compared to the conventional mode have been reported in the literature (3,10-13); this might be due to the differences in instrumental parameters in the different reports, including carrier flow rates, aspiration rates, and electronic integration. Some workers have experienced an inferior precision of 2 to 4% RSD working in the peak height mode with flow injection sample introduction both for atomic absorption and ICP spectrometry, either using the suction action of the nebulizer or pumping at a flow rate lower than the "natural" aspiration rate to propel the carrier solution (3,10,11). However, Brown and Růžička (13) have shown that when a peristaltic pump was used for propulsion and the carrier flow rate made *higher* than the natural aspiration rate, the precision for flow injection sample introduction could even surpass those obtained by conventional mode. Thus using this technique and copper as a model element, a precision of 0.44% RSD has been obtained by injecting 150 μL of an 8-ppm Cu solution compared to 0.78 to 0.89% for continuous aspiration, all without electronic integration for the atomic absorption measurement. As 80% of the steady-state signal could be reached with a 150-μL sample, the loss in sensitivity using flow injection sample introduction will be considered to be quite small when compared to the gain in precision and detection limits even better than continuous aspiration in the conventional mode might be expected. However, the use of electronic integration may improve the precision and detection limits using conventional sample introduction, whereas in the flow injection mode, because of the transient nature of the signal, the improvement will not be so pronounced; yet with careful control of the carrier flow rate and the natural aspiration rate of the nebulizer similar detection limits could possibly be achieved. Thus a carrier flow rate of at least 1 mL min^{-1} higher than the natural aspiration rate was recommended (13).

An important feature of flow injection sample introduction is the continuous aspiration of liquid phase (including carrier and sample) without the intermittent introduction of air into the nebulizer. While to what extent this feature improves the stability of an air-acetylene flame still remains to be evaluated, its beneficial effect on the induction coupled argon plasma is considered to be quite obvious (14,15). Introduction of air into the plasma during exchange of sample cups could be totally avoided with a flow injection system; this not only prevents the risk of extinguishing the plasma but also improves the long-term stability of the plasma. Jacintho et al. (14) using a flow injection ICP spectrometric system for the analysis of limestone obtained precisions of 1.34 and 1.23% RSD for calcium and magnesium, respectively, compared to 3.30 and 1.76% in the conventional mode. The improvement in precision has been attributed to the better stability of the plasma in the flow injection ICP system. The signals for calcium and magnesium (both 100 μg ml^{-1} concentration) varied only 1 and 0.7%, respectively, after 1 h of operation. Greenfield has reported a similar effect in improvement of stability of the plasma (15).

Relatively little information on the effect of mixing coil length upon reproducibility is available within the literature. Although Yoza et al. (16) have proposed a long mixing coil of 100 cm to improve the reproducibility of results obtained with flow injection sample introduction atomic absorption spectrometry, most workers prefer to use the shortest possible length from the injection valve to the nebulizer to limit the dispersion and increase the sensitivity. Good reproducibilities of less than 1% RSD have nevertheless been obtained in a number of cases using the shortest possible coil lengths with both atomic absorption and ICP spectrometric detection (2,13). It seems that the optimum coil length for good precision might depend on a number of other factors, such as the performance of the specific nebulizer, mode of propulsion, carrier flow rate, and the natural aspiration rate used in each case, and will not be very critical provided that the other factors are optimized.

2.4 Sample Throughput of Flow Injection Sample Introduction

In the simplest form of flow injection sample introduction, which does not involve separation, reagent addition, or other functions that increase the sample dispersion, the length of the mixing coil which connects the injection valve to the nebulizer of the spectrometer is usually a short length of narrow-bore (0.5 to 1 mm ID) plastic tubing. In such systems, very high sampling frequencies could be attained. For example, a sample throughput of 720 samples per hour has been reported by Rocks et al. (17) in the atomic absorption spectrometric determination of lithium in serum, injecting 10 μL of sample into a carrier solution pumped at 4 mL min^{-1} through a 25-cm mixing coil. A high sampling frequency of 514 samples per hour has also been reported in the analysis of copper, zinc, iron, and manganese in plants and soil extracts by atomic absorption (11). However, these high frequencies rather reflect the *capability* of the flow injection mode of sample introduction and not the *practicable* sample throughput which could be carried out in routine operations. In practice, enough time should be left to the operator between samples to change sample vessels, and if a syringe is used for filling the sample loop, to rinse the syringes. Even if an autosampler is used, one would need time to fill up and change the sample cups. Thus as reflected in most of the published work, usually lower sample throughputs of 120 to 240 samples per hour have been adopted in practice. As conventional atomic absorption or ICP spectrometry are fast analytical methods in themselves capable of analyzing over 100 samples per hour, the improvement in sample throughput which is practicable does not seem to be very impressive, and perhaps the real value of this form of sample introduction lies in its microsampling capability and other important features presented in other parts of this section.

2.5 Tolerance of High Salt Content in Samples

An important drawback of conventional sample introduction in flame atomic absorption or ICP spectrometry is the relatively low tolerance of salt content in the sample solution. It has been shown that continuous aspiration of concentrated solutions with 25 to 40% w/v salt content into an air-acetylene flame not only resulted in unstable signals but ultimately would extinguish the flame in a few minutes (6,18). These are the results from encrustation of solids on the nebulizer tips and burner slits. Special types of high solids nebulizers, such as the Babington or V-groove nebulizers have been designed to overcome this problem (1) and have been quite effective in preventing blockages in the nebulizer; however, the blocking of the burner slit or the sample introduction tube of a plasma torch could still occur and the problem still remains to be solved.

Flow injection sample introduction provides a most effective way of introducing samples with high solid contents. Mindel and Karlberg (18) successfully determined cadmium in a sample with 25% magnesium chloride. Fang et al. (6) analyzed copper in a 40% w/v solution of urea and potassium hydrogen phosphate fertilizer samples. In the latter case, 100 samples (80 μL sample volume) were injected successively at 15-s intervals without any detectable deterioration in the precision, and obviously the sampling process could have been carried on for much longer periods without trouble, whereas continuous aspiration in the conventional mode extinguished the flame in about 5 min. The highest tolerable salt content in the conventional sample introduction mode was tested and found to be about 10%. The reason for the improvement in performance might be due mainly to the short duration of sample transport through the nebulizer and burner. In a low-dispersion system the time required for a 100-μL sample bolus to pass through the detector may be less than 10 s; furthermore, owing to the dispersion of the sample, only the sample zone section corresponding to the peak maximum will have a concentration approaching the original salt content of the sample, thus restricting the exposure of the nebulizer and burner to high salt contents to only about 1 to 2 s for each injection. The small amount of solids that might have collected in the nebulizer-burner system during transport of the sample could readily be removed by the washing action of the carrier stream that follows, leaving little chance for accumulation. Direct introduction of samples with high salt contents might, of course, make background correction necessary; as almost all modern atomic absorption or ICP instruments have background correction as a standard function, this should not be considered as a serious drawback, especially when one considers the advantage that could be gained by increasing the sensitivity three- to fourfold (see Section 2.2) and saving of efforts in solution handling.

2.6 Viscosity Effects in Flow Injection Sample Introduction

Conventional atomic absorption and ICP emission spectrometric methods using the suction action of the pneumatic nebulizers for sample introduction suffer from variations in sample uptake rate with changes in sample viscosity. This effect can be minimized by using a pump to propel the sample at a definite flow rate. With flow injection sample introduction (with pump propulsion of carrier) this effect could be further minimized by dilution of the sample through mixing with the carrier. Tyson et al. (19) made an investigation of the viscosity effect in conventional and flow injection sample introduction by adding different proportions of glycerol to the aqueous sample solutions. Although a pump was used to propel the carrier, when large proportions of up to 50% v/v glycerol were added to the sample there was still a pronounced viscosity effect, producing more than 70% depression on the absorbance of 1 µg of magnesium per milliliter with steady-state measurements. With 200 µL of solution in the flow injection mode, even under low-dispersion conditions the depression effect of 50% glycerol decreased to about 40%, whereas with 30-µL injections the decrease in signal was only 10%, and no effect at all was observed below 20% v/v glycerol.

The effect of temperature of the sample solution on the absorbance signal has also been studied by the same group. A pronounced enhancement of about 30% was observed, with a rise in temperature from 25°C to 90°C when normal aspiration by suction was used. This effect was largely overcome by using a peristaltic pump to aspirate the sample; and if hot samples were injected into a carrier stream at room temperature, the effect on the signal was considered to be negligible, so that standards at room temperature may be used for calibration.

2.7 Various Techniques for Flow Injection Sample Introduction

Quite a number of modifications in flow injection sample introduction have been proposed by different workers to improve the performance of the technique, mainly in sensitivity and precision. Ito et al. (10), using an ICP emission system, proposed the injection of 40 µL of sample into a small air bubble in the carrier solution to limit the dispersion. The air segment was formed by lifting the carrier introduction tube out of the carrier solution for a short while. By doing so the peak response for boron has been increased almost twofold. Because the procedure was somewhat inconvenient to operate, the system was later automated (20). Despite the improvement in sensitivity, it should be considered, however, that the intermittent introduction of air into an induction coupled plasma may cause deterioration of system performance, resulting in a loss in precision which otherwise would have improved by using the continuous mode of liquid introduction (see Section 2.3).

Some workers have used organic solvents as a carrier for aqueous samples or/and as sample solvents to enhance the sensitivity and lower the detection limits with atomic absorption detection (see Section 2.2). Methanol, ethanol, acetone, MIBK, and n-butyl acetone have been tested for their enhancement effects (8,9). It is assumed that solvents immiscible with water produced enhancements due mainly to limitations in the dispersion of the sample plug, whereas the enhancement effects of the water-miscible solvents might be due principally to increases in atomization efficiencies in a higher-temperature flame in the presence of a suitable organic solvent. Combinations of organic carriers with immiscible or partially miscible organic sample solvents which exhibited the highest enhancements probably produced both effects.

Brown and Růžička (13) have proposed augmenting the carrier flow with an additional confluence stream to retain a higher flow rate than the natural aspiration rate when the carrier flow rate is lower or equal to the latter flow rate. This was considered to be necessary to maintain good precision. The loss of sensitivity due to dilution from the extra flow was found to be small; the reason was thought to be an improvement in the nebulization efficiency under such conditions. With a similar purpose Yoza et al. (16) have proposed augmenting the carrier stream by aspirating the extra flow through a T-piece immediately before the inlet of the nebulizer through the suction created by atmospheric pressure if the carrier stream flow rate was lower than the natural aspiration rate. They also recommended the compensation for the difference in flow rate by introducing air through the T-piece. Higher sensitivity was obtained because of the absence of dilution effects from the extra stream, but precision was found to be somewhat poorer.

3   DILUTION TECHNIQUES

In atomic absorption and ICP emission spectrometric determinations sample solutions often have to be diluted before measurement either because of limitations in the linear range of the calibration curve (especially in atomic absorption methods) or due to various kinds of interference effects from the matrix which could be suppressed through dilution. Dilutions could readily be performed automatically in flow injection systems, considerably simplifying the volumetric manipulation in routine analysis. The various modes for flow injection sample dilution suitable for atomic spectrometric analysis are reviewed in this section.

3.1 Dilution by Simple Dispersion Manipulation

As controlled dispersion is considered to be one of the cornerstones of flow injection analysis, it is a relatively easy matter to control the degree of dilution or dispersion of an injected sample by optimizing the sample volume, carrier flow rate, and mixing coil lengths in any flow injection sample introduction system. Although this is the most convenient and simplest way to dilute samples, its application is limited in that when extremely small volumes (a few microliters or less) are used to obtain large dispersions and high dilution factors, the precision might be reduced. On the other hand, if long mixing coils are used to increase dispersion, the sampling frequency will be lowered. Hence it might be wise to limit the dispersion coefficient (or dilution factor) to less than 100 when using this mode of dilution.

The potentials of dilution through dispersion in single-line manifolds have been well illustrated by the works of Rocks et al. in clinical analysis applications (4,17). In the analysis of serum samples by atomic absorption spectrometry a dispersion coefficient of 12 was obtained by injecting 10-µL samples into a system with a 25-cm mixing coil (4); the injection of 4-µL samples into a system with a 200-cm coil produced a dispersion coefficient of 54, whereas increasing the tube diameter from 0.8 mm to 1 mm with the same coil length increased the coefficient to 100 (17).

3.2 Dilution by the Merging of Sample and Diluent

Dilution of sample could also be accomplished by merging the carrier stream containing the dispersed sample with a diluent stream as shown in Fig. 2. Zagatto et al. (21) achieved a 40-fold dilution of the sample in the determination of calcium and magnesium by atomic absorption, and potassium by emission spectroscopy by pumping the carrier at 1 mL min^{-1} and the diluent at 12 mL min^{-1}, and injecting 10 µL of sample into a manifold with coil lengths of 230 cm for magnesium and potassium and 99 cm for calcium. The method was capable of analyzing 300 samples per hour, with a typical relative standard deviation of 0.5%. This approach for on-line dilution is quite straightforward and easily applied; however,

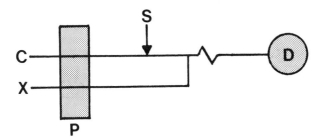

FIGURE 2 Merging stream flow injection dilution manifold. S, sample; C, sample carrier; X, diluent; P, pump; D, detector. [Based on (21) with permission of Elsevier Science Publishing.]

extremely large differences in the carrier and diluent flow rates might be impractical and might result in degradation of the precision.

3.3 Split-Flow Dilution

Mindel and Karlberg (18) have proposed a dilution system for atomic absorption spectrometry which involved splitting the flow stream containing the dispersed sample by drawing off a certain proportion of the stream at point F in Fig. 3. The loss in flow rate was compensated for by introducing diluent at point E at the appropriate flow rate. In such systems the dilution factor depends on the proportion of flow which is drawn off and made up downstream, and could be altered by using different draw-off rates. Although the manifold is more complicated, versatility is gained, as dilution factors could be varied without changing the sample volume or mixing coil length.

3.4 Dilution by Zone Sampling

The zone sampling technique introduced by Reis et al. (22) is capable of effecting large dilutions efficiently with dispersion coefficients of over 100 and with good precision. In a zone sampling system the injected sample is dispersed in the carrier stream, and instead of flowing directly into the nebulizer, the sample zone is introduced into a second sample loop on the same injection valve (commutator). The capacity of the second loop is chosen so that only a small portion of the dispersed sample zone could be held in it (Fig. 4). Thus, by precise timing of the actuation of the valve, a predetermined "slice" of the dispersed sample zone is sampled for a second time when the valve is turned back and dispersed again in a second carrier stream before entering the nebulizer. Reis et al. (22) reported a 130-fold dilution in the determination of

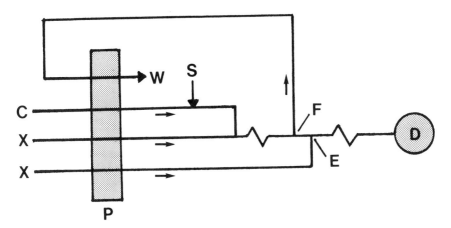

FIGURE 3 Split-flow injection dilution manifold. S, sample; D, atomic spectroscopic detector; P, pump; W, waste; F and E, points for draw-off and compensation streams; C, carrier; X, diluent. [Based on (18) with permission of Elsevier Science Publishing.]

potassium in plant digests with atomic absorption spectrometry at 120 samples per hour. Good precision was also achieved but no values were given. Although specially designed valves and equipment for precise timing are necessary to carry out the zone-sampling process, such systems do exhibit a high flexibility in controlling the extent of dilution, since the dilution factor may be changed simply by altering the time lapse between the first and second injections. This feature should be very useful when the concentration

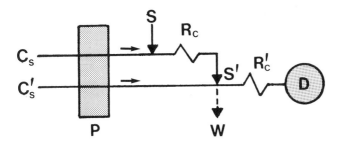

FIGURE 4 Schematic diagram of zone-sampling flow injection manifold. C_s and C_s', carrier streams; R_c and R_c', mixing coils; D, detector; S, sample; S', "slice" of dispersed sample; W, waste; P, pump. [Reproduced with permission of Elsevier Science Publishing (22).]

range of the analyte is wide and different dilution factors are desired within one batch of samples.

4 TECHNIQUES FOR ADDITION OF REAGENTS

In atomic absorption determinations, certain reagents sometimes have to be added to the samples as releasing agents or buffers. The volumetric operations involved often take up a considerable percentage of the total time consumed to make an analysis. These operations can readily be accomplished on-line by using various designs of flow injection manifolds for reagent addition.

4.1 Addition by Using Reagent as Carrier

Reagents may be added to the sample simply by using the reagent solution as a carrier stream. The sample zone is dispersed into the reagent carrier in a very reproducible manner, producing a certain degree of mixing. The extent of mixing will depend on the dispersion, the sample/reagent ratio decreasing with an increase in the dispersion coefficient of the flow injection system. A flow injection system with a large dispersion will almost retain in the dispersed sample zone the original concentration of the reagent carrier, whereas in a low-dispersion system hardly any reagent will be found at the sample peak. Hence this approach for reagent addition will be suitable mainly for systems with dispersion coefficients of 2 (corresponding to a sample/reagent ratio of 1) or more. Rocks et al. used a carrier stream of 10^{-2} M lanthanum in the determination of calcium and magnesium in serum by atomic absorption with a large-dispersion flow injection system (D_t = 54) (17). In this case the reagent concentration at the sample peak would be almost identical to that of the carrier solution. The same group has also used 2×10^{-2} M EDTA as carrier in a system that had an even larger dispersion (D_t = 100). EDTA was added to the sample to overcome interferences in the determination of calcium and magnesium.

Although the continuous feeding of reagents in the form of a carrier might appear rather wasteful because actually only the central section of the sample zone needed the addition of reagents, the consumption of reagents is usually an order of magnitude lower than that of conventional aspiration. In the case of lanthanum this will amount to only a few milligrams of reagent for each sample as compared to about 50 mg per sample in the conventional mode of operation.

4.2   Reagent Addition by Flow Confluence

In this mode of reagent addition the sample is injected into a carrier stream while the reagent is pumped through a separate line;
the two streams merge and mix at a confluence point downstream.
The same type of manifold used for dilution shown in Fig. 2 may
be used for this purpose. If higher sensitivity is desired, the addition of reagents could be achieved without creating excessive dispersion by using low reagent flow rates with higher reagent concentrations, thus limiting the dilution of the sample to a minimum.
Basson and van Staden (23) used such a system to add a reagent
solution containing lanthanum, cesium, and lithium to an aqueous
carrier stream (both at 2.5 mL min^{-1} flow rate) in the simultaneous
determination of sodium, potassium, magnesium, and calcium by
flame photometry and atomic absorption spectrometry.

4.3   Merging-Zone Reagent Addition

The foregoing techniques for reagent addition are all based on continuous feeding of reagent solutions, which implies that reagents
are also present in that part of the stream that contains no sample.
Even so, the consumption of reagents is much lower in the flow injection mode compared to conventional operation.   Nevertheless,
when expensive reagents are used, a more economical approach
would be desired.   In the merging-zone method, sample and reagent are introduced into two separate loops often arranged on a
dual-channel valve and injected simultaneously into two parallel carrier streams (Fig. 5).   The flow rates and tube dimensions are

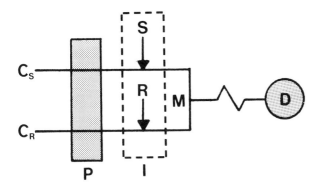

FIGURE 5   Merging-zone flow injection manifold for reagent addition.
C_S, sample carrier; C_R, reagent carrier; M, confluence point of
merging zones; D, detector; S, sample; R, reagent; I, injector;
P, pump. [Based on (21) with permission of Elsevier Science
Publishing.]

chosen so that after injection the sample and reagent zones within the two streams merge at a certain point downstream. The reagent consumption is thus reduced considerably. Using this technique, Zagatto et al. (21) were able to reduce the consumption of lanthanum to 0.5 mg per sample in the determination of calcium and magnesium in plant materials by atomic absorption, corresponding to about 1% of the amount consumed in conventional operation. This has also been verified in the author's laboratory (11).

5 CALIBRATION METHODS

Flow injection peaks not only provide reproducible readouts at the peak maxima, but also contain a wealth of information on analyte concentration on the rising and falling sections of the peak gradient, as well as information from the peak area and peak width. Flow injection also provides efficient techniques on sample dilution and reagent addition, as has been shown in the preceding sections (see Sections 3 and 4). These important features have been exploited extensively to provide more efficient calibration methods in addition to the conventional approach of plotting a working curve with a standard series. In this respect, atomic spectroscopy seems to be a field on which special interest has been focused. Various techniques and methods have been developed to perform calibration with atomic spectrometric systems, and several are discussed in this section.

5.1 Calibration Through Variation of Dispersion of Standard

As has been shown in Section 3.1, the dispersion or dilution factor of a standard sample could be varied by injecting different volumes of the same standard solution and/or by using different mixing coil lengths. Thus, in principle, a calibration curve could be constructed by using a single standard solution. Although experimentally feasible, the creation of different dilution factors by changing sample loops with different volumes does not show much advantage over conventional serial dilution operation for the construction of a calibration curve, but the switching of the injected volume sequentially down into a set of parallel conduits with different tube dimensions to give different dilution factors seems to be simple and rapid. A versatile dilution system has been constructed by Tyson et al. (19) by inserting two six-way stream-switching valves in the manifold and connecting the various inlets and outlets of the valves with different lengths of 1.58-mm-ID tubing (110, 250, 415, 560, 1115, and 2000 mm, corresponding to D_t values of 2.52, 3.49, 4.43, 5.24, 7.53, and 14.9, respectively). Sixty-five microliters of a standard solution whose concentration was higher than the normal working range was injected sequentially into the six lines and the peaks recorded. The dilution factor had an uncertainty of 1 to 2%, which is generally acceptable; and the longest time for a peak to return to baseline was 90 s.

5.2 Calibration by Peak Width Measurements

When a low dispersion system is used for sample introduction in atomic absorption spectrometry, the nebulizer is the dominant factor in determining the peak shape, and approximately exponential shaped peaks are to be expected. In such cases the width of the peak will be proportional to a logarithmic function of the concentration of the analyte (24). If the peak width is plotted against $\ln[(C_m/C') - 1]$, where C_m is the concentration of the analyte and C' is the concentration level chosen for the peak width measurement, a straight calibration curve is obtained. Alternatively, the peak width could simply be plotted against $\ln C_m$ with some sacrifice in linearity, which implies that strict adherence to exponential peak shapes might also not be necessary. Thus, if a solution is too concentrated to be measured, either because of excessive loss of linearity or creating an off-scale signal, the peak width mode of measurement could be applied. The logarithmic relation has been verified experimentally over the range 1 to 1000 μg of magnesium per milliliter with an uncertainty of about ±5% relative over the complete concentration range, which is considerably higher than that which could be obtained in the conventional peak height mode. The method might prove useful in providing guidelines for appropriate dilution factors for subsequent accurate analysis or in situations where a large concentration range has to be dealt with but utmost precision is not required.

5.3 Calibration by the Exponential Flask Method

The exponential flask calibration method is based on the use of the well-defined exponential concentration-time gradient generated when a step concentration change input passes through a well-stirred tank mixing chamber. The latter is connected with the nebulizer of an atomic absorption or ICP spectrometer, and the effluent from the mixing chamber nebulized in the conventional manner. The concentration-time relationship of the effluent satisfies the following expression:

$$C = C_s[1 - \exp(-ut/V)]$$

where

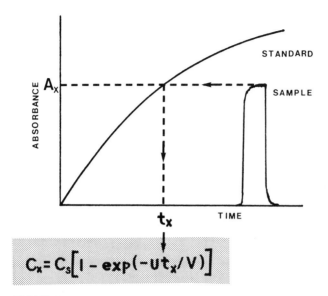

FIGURE 6   Absorbance-time curve obtained with exponential flask
method.   [Reproduced with permission of Elsevier Science Publishing (25).]

C_S = original concentration of the standard solution introduced
into the tank

C = concentration at time t measured from the beginning of
the step change

V = volume of the mixing chamber

u = flow rate

If we assume a linear relationship between instrument response A
and concentration of sample, the corresponding A-t curve (Fig. 6)
is given by

$$A = A_s[1 - \exp(-ut/V)]$$

where A_s is the steady-state response.   The response A_x of an
unknown sample aspirated in the conventional way at the same flow
rate is used to obtain t_x from the curve, which in turn is used to
deduce C_x from the concentration-time relationship.   Tyson (26)
and Greenfield (27), using atomic absorption and ICP systems, respectively, showed that the experimental behavior of a concentration-time profile was in good agreement with the predictions.

5.4 Calibration by Interpolative Standard Addition

In conventional standard addition procedures, calibration is usually achieved by extrapolation of the standard addition points. The interpolative standard addition method proposed by Tyson et al. (28) is characterized by interpolation between positive and negative values obtained by injecting a set of *standards* into an unknown *sample* carrier stream. Standards with a higher concentration than the sample produce positive peaks on the sample carrier baseline, whereas those with a lower concentration give rise to negative peaks. Provided that interference effects are kept within certain limits, identical concentrations of samples and standards are not expected to produce peaks. When the concentration of the standards (C_S) are plotted versus peak heights (ΔA) the intercept on the C_X axis corresponds to the concentration of the unknown sample (Fig. 7). The interpolative method has certain advantages over the extrapolative mode, in that linearity of the concentration versus peak response relationship is not necessary, and better compensation of spectral interferences might also be possible if the dispersion of the injected standards is carefully controlled. The dispersion of

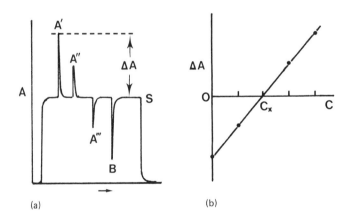

(a)   (b)

FIGURE 7   (a) Typical recordings of responses for an interpolative standard addition method. ΔA, difference between peak response and sample carrier response; S, steady-state response of sample carrier; A', A", and A"', peaks for different levels of standard additions; B, blank. (b) Plot of results from part (a) for the evaluation of results by interpolative standard addition method; ΔA versus C, concentration of standard; when $\Delta A = 0$, $C = C_X$, concentration of sample. [Reproduced with permission of Elsevier Science Publishing (28).]

the standard should be chosen so that the ratio of sample to standard at the peak maximum where readouts are taken will not be too
low to ensure adequate compensation of interferences, yet excessive
dilution of the standards will lower the sensitivity and hence the
precision of the measurements. Through careful control of the concentration of the iron matrix interferent, Tyson et al. (28) obtained
excellent results in the analysis of chromium in steels by the method
proposed. However, interference control will be more difficult in
such samples as rocks, soils, or plant materials, where the sample
matrices are more varied (6). In such cases better control over
interferences might be achieved by using the interpolative method
with an ICP emission system.

The interpolative standard addition method has been used successfully by Greenfield (27), Israel and Barnes (29), and Fang et
al. (6) with ICP emission systems. This method could be advantageous in minimizing errors arising from source drift or wavelength
drift. Another advantage is that the wide dynamic range of the
ICP source allows further simplification of the method. Thus, by
assuming a linear intensity-concentration relationship, the following
equation for standard addition has been deduced by Israel and
Barnes (29):

$$C_x^o = \frac{I^m C_s^o}{I^m + I^p}$$

where C_x^o and C_s^o are the concentrations of the sample and standard solutions added; I^m and I^p are the peak heights of the blank
and standard relative to the sample baseline response I_x^o. It could
be anticipated from the equation above and considering the signal-
to-noise ratio of the measurement that I^m/I_x^o and I^p/I_x^o should be
kept high to increase the precision of the measurement of I^m and
I^p. Thus large injection volumes for the standard, corresponding
to a dispersion coefficient of up to 4 for the sample and a C_x^o/C_x^o
ratio between 2 and 2.5 were recommended. The method is simple
in that only two injections are necessary for a determination. However, owing to the low dispersion of the standard, the matrix concentration at the peak maxima will be quite different from that of
the original sample, which gives rise to the baseline signal. Therefore, the method is not expected to compensate for spectral or chemical interferences but will be effective in obviating short- and long-
term instabilities. The method was tested by Fang et al. (6) using
some standard reference materials and following the criteria on sample and standard dispersion as recommended by Israel and Barnes
(29). Satisfactory results were obtained for barium and chromium,

showing good agreements with certified values (6); yet when chemical or spectral interferences are more pronounced, a dispersion
coefficient of approximately 2 (i.e., a 1:1 dilution of sample to
standard at the peak maximum) might provide a better compensation of the interferences at the expense of some loss in precision.

5.5 Calibration by Zone Sampling

The zone sampling technique introduced by Reis et al., which has
been used for on-line dilution (see Section 3.4), has also been
adapted to provide a simple calibration method (22). By varying
the time interval between the first and second injections, a different "slice" of the dispersed *standard* zone could be chosen for each
standard injection, each time corresponding to a different dispersion
coefficient or standard concentration. When a time interval-disper-
sion coefficient relation is set up, a complete calibration graph
could be constructed by using a single standard. However, no
actual applications with an atomic spectrometric system seems to
have been reported thus far.

5.6 Merging-Zone Generalized Standard Addition
Method

The merging-zone approach for reagent addition (see Section 4.3)
could be readily adapted to the performance of calibration by standard additions. Different standards are added to the samples directly, analogous to conventional standard addition procedures, but
volumetric manipulation is considerably simplified. Extra work will
be involved to synchronize the merging of the zones, and the dispersion coefficients of the merging peaks will also have to be estimated, but these will be outweighed when a large number of samples have to be analyzed using the standard addition technique.
    The generalized standard addition method (GSAM) proposed by
Saxberg and Kowalski (30) is an effective calibration method to overcome interelement interferences; yet the method involves considerable volumetric manipulation, normally requiring a number of additions
of standard solutions to the sample in order to estimate the mutual
interferences and matrix effects. The flow injection merging-zone
approach could significantly simplify the manipulations, with large
savings in time and sample material. The method has been applied
to the analysis of copper-nickel alloys by ICP emission spectroscopy
by Zagatto et al. (31) to overcome the spectral interferences and
other matrix effects in the determination. The results obtained
were in good agreement with those obtained by the manual procedure.

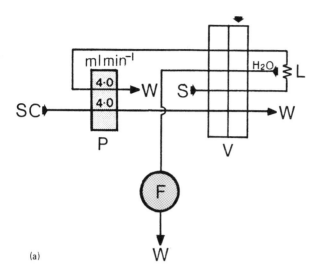

(a)

FIGURE 8 Single-injection standard addition method. (a) Manifold; V, valve; L, sample loop; F, flame photometer; S, sample;
SC, standard solution carrier; P, pump; W, waste. (b) Time-
intensity recordings of an 80-mg 1^{-1} standard calcium solution injected as a 100-μL sample with water as carrier to show the dispersion of the sample zone A. The dashed lines indicate the times
and dispersions when identical D values can be obtained on the rising and falling sections of the peak. At the top is a schematic
diagram of the conditions in the flow channel immediately prior to
nebulization. B is a recording of the same standard used as carrier and distilled water injected as sample. Note that the standard
solution is not dispersed into the rising part of the sample zone.
W, water; S, sample; SC, standard carrier; t_1 and t_2, data sampling times; D_1, D_2, and D_3, dispersion coefficients for points of
identical D values. (c) Superimposed recordings of 100-μL soil
extract samples with 80 μg of calcium per milliliter as carrier.
Curves 1, 2, and 3 are soils with different levels of calcium concentration. H_r and H_f are the responses measured at t_1 and t_2 at
identical dispersions of the sample zone for curve 2. [Reproduced
with permission of the American Chemical Society (32).]

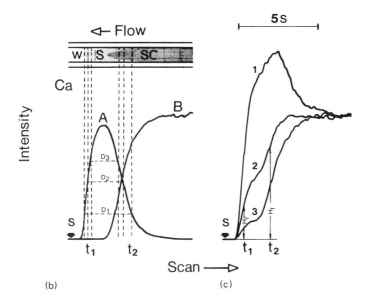

(b)   (c)

5.7 Single-Injection Standard Addition Methods

A simple standard addition method involving only one injection for
each determination has been proposed by Fang et al. (32). Although
the method was used in combination with the gradient scanning technique for flame photometric simultaneous multielement analysis, it
could be equally well adapted to single- or multielement analysis by
atomic absorption or ICP spectrometry. The flow injection manifold
for making the standard addition is shown in Fig. 8(a) and is arranged so that when a sample is injected in the normal way, the
carrier upstream is water or the blank solution and the carrier
downstream is a standard solution of the analyte, with the sample
sandwiched in between. The manifold is designed to produce a low-
dispersion system with the dispersion coefficient close to unity to
ensure that no overlapping of the two carriers occurs. On the falling gradient of the sample, this arrangement could produce an almost infinite number of sample/standard ratios. The ratios at different delay times after injection could easily be computed from dispersion measurements. An optimum ratio or several ratios within
the optimum range are chosen by setting the appropriate delay time
t_2 or a set of t_2 values to perform data sampling and storage during
each injection. If the system is used for multielement analysis, optimum ratios at different t_2 values could be chosen for each individual analyte to perform the measurement of the signals. For any

t_2 taken for measurement, a corresponding t_1 is defined on the rising gradient of the sample peak, which represents a point having the same dispersion as the point at t_2, yet without addition of standard [Fig. 8(a)]. If we assume a linear concentration-intensity or absorbance relationship, the analytical signals H_r and H_f on the rising and falling sections of the peak obtained at a pair of t_1, t_2 are related to sample concentration C_{sm} according to the equation

$$C_{sm} = K \frac{H_r}{H_f - H_r}$$

for a certain dispersion D_{sm} and concentration of the standard C_{st}, K is a constant

$$K = C_{st} \frac{1 - d}{d}$$

where $d = 1/D_{sm}$. The method has been applied successfully to the determination of calcium in soil extracts by flame photometry, but its full potential might be explored only when used with a multichannel fast-scanning spectrometric system.

Another single-injection standard addition calibration method was proposed by Araújo et al. (33). Like the zone sampling calibration technique, the method is based on the exploitation of the concentration gradient of an injected and dispersed standard solution. However, the procedure is greatly simplified by injecting the standard into a carrier stream and defining a series of concentrations at predetermined time intervals. The dispersed standard is subsequently merged with the undispersed sample to achieve a simple standard addition procedure performed under constant matrix conditions. The flow system and recordings of the standard addition method are shown in Fig. 9. The standard solution is initially pumped through the *sample* line to evaluate H_p (response corresponding to concentration of original standard). Then, by merging the dispersed standard with a blank in the sample stream, the heights H_1, H_2, ..., H_n on the falling gradient of the standard peak, associated with appropriate standard addition concentration levels, are selected and the corresponding time intervals following injection are evaluated manually. H_1', H_2', H_3', ..., H_n' for all samples are then measured at the same time intervals. Analyte concentration is deduced from the two sets of time-dependent height measurements by linear regression.

The method has been tested with samples containing 0 to 10% (v/v) ethanol in the determination of calcium, potassium, and sodium by flame photometry. Errors were less than 3%, whereas the

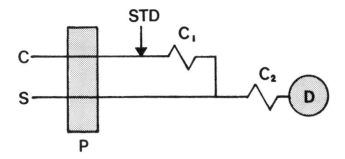

(a)

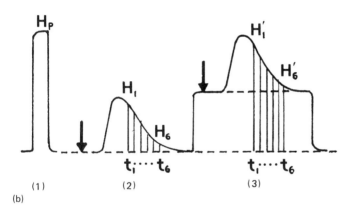

(b)

FIGURE 9 (a) Single-injection standard addition manifold. C_1 and C_2, mixing coils; S, sample; STD, standard; C, carrier; P, pump; D, detector. (b) Recordings obtained with the manifold in (a). (1) Standard solution pumped continuously in the sample line; (2) same standard solution injected into water carrier and merging with blank solution in sample stream; (3) same standard solution injected into water carrier and merging with a continuous flow of sample solution. $t_1 \cdots t_6$, time intervals when response measurements $H_1 \cdots H_6$ are taken for recording (2) and $H_1' \cdots H_6'$ are taken for recording (3). [Reproduced with permission of Elsevier Science Publishing (33).]

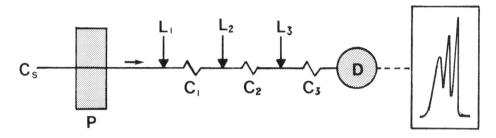

FIGURE 10 Schematic presentation of the sequential injection process. L_1, L_2, and L_3, sampling loops arranged on the same injector; C_1, C_2, and C_3, mixing coils; C_s, carrier stream; P, pump; D, detector. [Reproduced with permission of Elsevier Science Publishing (35).]

conventional calibration graph method produced 42% error for sodium when an aqueous standard was used to analyze a sample with 10% ethanol.

5.8 Calibration by Sequential Injection

The precision of calibration methods based on the exploitation of concentration gradients such as those described in Sections 5.3, 5.4, and 5.7 all rely on the reproducibility of the concentration-time profile of the injected sample or standard and on the precision of the timing. The precision of such measurements will be degraded if a fast-changing gradient must be used to provide the necessary readout. The sequential injection technique proposed by Hansen et al. (34) has been used by Zagatto et al. as an alternative for gradient exploitation to improve the precision of measurements and has been used successfully for calibration purposes with atomic absorption spectrometry (35). The technique involved the simultaneous intercalation of several plugs of analyte solution into the same carrier stream (Fig. 10). Upon dispersion, the plug zones overlap each other to a certain extent downstream as a result of asynchronous merging. A multipeak response is obtained, characterized by several maxima and minima sites which are almost without gradients. Only these points are taken for measurements because they are less affected by fluctuations in the flow parameters. If a standard is taken for sequential injections, the dispersion of the plugs could be devised so that the maxima and minima of the concentration-time profile represent a broad range of dilution factors of the standard solution. A standard series run could provide several calibration curves at different concentration levels. An unknown sample using sequential injection will produce sites of measurements without

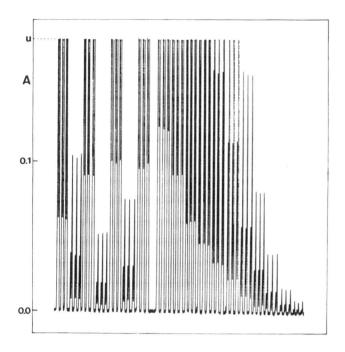

FIGURE 11   Recordings for the atomic absorption determination of manganese in rocks by sequential injection of two sample zones. From right to left: 11 standards (0.10, 0.20, 0.50, 1.00, 2.00, 4.00, 6.00, 8.00, 10.0, 15.0, and 20,0 μg mL^{-1} manganese) followed by seven samples, all in triplicate.   The upper recording threshold (u) corresponds to the scale boundary of the recorder above which pronounced loss in linearity occurs.   [Reproduced with permission of Elsevier Science Publishing (35).]

gradients and with the same dilution factors as those of the standards.   The optimum dilution factor could then be chosen to evaluate the results.   The recordings in Fig. 11, obtained in the analysis of manganese in rocks, show the broad range of concentrations that could be manipulated in the calibration with good precision.   The sampling rate was 100 samples per hour, and the results were in good agreement with certified values of standard rock samples.   The technique has been used by the same group to perform standard additions in the determination of copper in ethanol.   In this application the overlapping plug zones of a standard solution are merged with a sample solution stream which has attained a steady-state signal.   The different maxima and minima in the resulting response then correspond to the addition of different concentrations of the

standard solution, which could easily be computed by comparison
with steady-state signals of the standards.

6   INDIRECT METHODS FOR ATOMIC ABSORPTION
SPECTROSCOPY

Considerable work has been undertaken toward the development of
indirect methods of improving the sensitivity of certain species in
atomic absorption determinations. Applications of flow injection
techniques to such methods have been reported to improve the selectivity of the procedures as well as improving sample throughput
and sample/reagent savings compared to conventional methods based
on similar principles; it might not be possible at all to perform
some of the applications in the conventional mode (see Section 6.2).

6.1   Indirect Methods Based on Enhancement Effects

It is known that certain elements are capable of enhancing the absorption signal of iron in a fuel-rich air-acetylene flame (36).
Martínez-Jiménez et al. (37) developed a simple flow injection procedure based on this principle for the indirect determination of
aluminium. Fifty-four microliters of sample is injected into an
iron(III)-tartaric acid carrier stream (pH 3.5 to 6.0) which is
pumped at 3.0 mL min^{-1} through a 100-cm mixing coil. The iron
signal of the carrier is taken as the baseline and the peak resulting from the aluminum enhancement is correlated with aluminum
concentration in the range 0.2 to 1.8 µg mL^{-1}. A high sampling
frequency of 150 samples per hour has been achieved with large
improvements in selectivity. The tolerance of coexisting fluoride
in the sample was reported to be two orders of magnitude higher
than the conventional indirect method, and the tolerance of lead
and bromide was one order of magnitude higher, but no explana-
tions were given for the improvement. Small amounts of calcium
still interfere in the determination but could be separated by an
oxalate precipitation. A similar scheme has been proposed by the
same group for the determination of uranium, lanthanum, and cerium with a sensitivity 2000- to 5000-fold higher than the direct
method using a nitrous oxide acetylene flame (38,39). For lanthanum and cerium an iron(III)-tartrate carrier solution similar to the
composition used for the determination of aluminum was used, but
for uranium the iron(III)-EDTA carrier was found to yield better
precision and selectivity. The indirect methods for lanthanum,
cerium, and uranium show considerable interference from common
ions and hence are suitable only for relatively pure analyte solu-

tions or for analyses where separation is used to remove potential interferents.

6.2 Indirect Methods Using Conversion Techniques

Anions that cannot be determined by flame AAS could be converted into a species containing a "tag material" which can readily be determined at a subsequent stage. Flow injection techniques make on-line conversion feasible, and conversion into both soluble and insoluble species has been developed. Haj-Hussein et al. (40) reported an indirect method for the determination of cyanide by atomic absorption through the formation of cupro-cyanide complex. Aqueous samples (pH 10 to 11) were injected into a potassium hydroxide carrier solution and passed through an on-line microcolumn packed with cupric sulfide. Cyanide ions in the sample were transformed into cupro-cyanide complex and the stream was subsequently introduced into the nebulizer of the spectrometer, where the copper concentration was determined. A detection limit of 1 μg mL^{-1} for cyanide has been achieved with a sampling frequency of 40 to 50 samples per hour.

Petersson et al. (41) developed an indirect method for the determination of sulfide by its conversion into the insoluble cadmium sulfide precipitate, using cadmium sulfate as the tag reagent. Two hundred microliters of sample was injected into an ammonium acetate buffer carrier (pH 8) and mixed with 250 μL of cadmium sulfate reagent using the merging-zone technique. Excess of cadmium was retained on an ion-exchange column packed with a chelating resin, while the cadmium sulfide precipitate, which was presumed to be primarily in the colloidal form, passed through the column and was then introduced into the nebulizer. Sulfide was determined by measuring the cadmium signal; subsequently, the absorbed cadmium was eluted by 1 M HNO_3. Thus for each analysis cycle, two peaks were recorded, one rising from the transport of cadmium sulfide into the detector and the other due to the elution of excess cadmium held on the column (Fig. 12). The detection limit for sulfide was 10 μg L^{-1}. A sampling frequency of 100 samples per hour could be attained using a semiautomated system. Among the potential interfering species studied, only phosphate had a positive effect on the analytical signal. It is interesting that no interference was observed when phosphate but no sulfide was present in the sample. This phenomenon was thought to be due to a coprecipitation effect during the formation of cadmium sulfide. The interference was overcome by adding 200 μg of phosphate per milliliter to all samples and standards, so that a further increase in phosphate concentration of 50 μg mL^{-1} would produce only a 5% enhancement in signal.

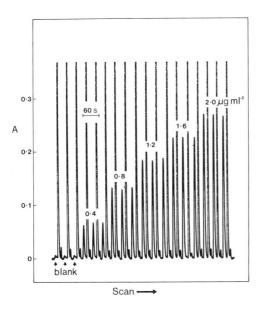

FIGURE 12   Recordings for the determination of sulfide by conversion flow injection method with atomic absorption detection.   Calibration run in the range 0 to 2 µg mL^{-1}.   The peaks are obtained from atomic absorption detection of the cadmium sulfide formed. The elution peaks of the adsorbed cadmium are all out of range of the chart recorder, so they appear to be of equal heights.   [Reproduced with permission of Elsevier Science Publishing (41).]

7   HYDRIDE GENERATION METHODS

Hydride generation coupled with atomic absorption or ICP spectrometric detection has been used extensively for the determination of arsenic, selenium, bismuth, antimony, and other metals in various matrices.   The hydride-forming reaction is usually processed in a reactor using sodium or potassium borohydride as the reductant in acid medium; the hydride and hydrogen formed are swept into the atomizer with a suitable carrier gas.   The determination is usually fast—only 1 to 2 min to obtain the analytical readout—but sample consumption is relatively large compared to that of other atomic absorption or ICP emission methods.   Interference might also be serious in some cases, and considerable research has been carried out to improve the tolerance of various interferents.   Applications of the flow injection technique to hydride generation atomic spectroscopy have produced very efficient methods giving excellent precision, with sample and reagent consumption more than one order of magnitude lower than that using conventional procedures and often with less interference.

7.1 Hydride Generation Flow Injection Manifolds

The first flow injection hydride generation atomic absorption system was developed by Åström, who used it to determine bismuth at µg L^{-1} has been achieved with a 700-µL sample at a sampling frequency of 180 samples per hour. Precision was excellent, but no data for the carrier flow rate, which is quite important for ensuring good precision, were given in the paper. It is quite interesting to see how closely Åström's setup matches that contemplated by Greenfield (27). Since Åström's paper, several articles on the flow injection hydride generation determination of selenium, arsenic, lead, antimony, and tellurium in different sample matrices have appeared (43-47). The manifolds are generally quite similar; the one used by Wang and Fang (6,43,44) for the determination of selenium is shown in Fig. 13. These manifolds are usually characterized by much larger carrier flow rates and sample volumes than those of other flow injection systems. This might be due to the larger dead volume of certain components in the manifold, including the gas-liquid separator and the quartz absorption cell, which is the most commonly used form of atomizer.

Another characteristic is that a stripping gas, usually nitrogen or argon, introduced at a flow rate of approximately 100 to 200 mL min^{-1}, is always used in the manifold to assist in liberation of the hydride. Control of the flow rate of the stripping gas is quite important. Although the flow rate is not considered to be very critical, it should not fluctuate excessively when large amounts of hydrogen are liberated in the system, thus generating considerable pressure in the narrow tube system. In the author's laboratory a throttle in the form of a small copper disk with a tiny aperture of about 0.2 mm was fixed in the outlet of the pressure gauge of the gas cylinder. A back pressure of more than 3 kg/cm^2 could be produced when the gas flow was as low as 150 mL min^{-1}, and the flow rate was not affected by small back-pressure variations in the flow system, thus ensuring a uniform flow.

Incorporation of a gas-liquid separator is another common feature of all hydride generation flow injection systems. The separator is necessary for separation of the hydride and other gases from the aqueous reaction waste, and in most cases a Vijan-type U-tube gas-liquid separator (48) was used for this purpose. Owing to the smaller sample volume often used in flow injection systems compared to that in segmented continuous flow systems or other hydride generation systems, the dead volume of the gas-liquid separator would be expected to be more critical. Smaller dead volumes will no doubt

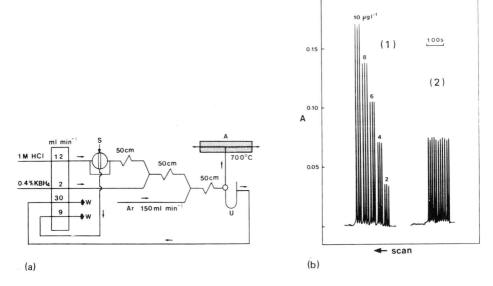

(a)   (b)

FIGURE 13   (a)   Manifold for the hydride generation atomic absorption spectroscopic system.   U, gas-liquid separator; S, sample, 400
µL; W, waste; A, T-shaped quartz atomizer with electrical heating.
(b)   Recordings for standard selenium solutions obtained with the
system in part (a).   (1)   Standard series in the range of 2 to 10
µg L^{-1}; (2) replicate injections of a 4-µg L^{-1} selenium standard.
[Reproduced with permission of Elsevier Science Publishing (6).]

be favorable in improving sensitivity by limiting the dispersion of
the gas phase, but will bring more risks in incomplete gas-liquid
separation.   As entrainment of liquid phase into the atomizer could
produce disastrous results, including bad precision, contamination
problems, and shortening the lifetime of the quartz atomizer, the
effluent flow is drawn to waste by a separate pump line.   A drawn-
out flow rate which is higher than the total in-flow rate of the
liquid phase is always used to ensure that no liquid enters the
atomizing system.   In the author's laboratory, the dimensions of
the expanded bulb of the U-tube gas-liquid separator were made
smaller than those of the original Vijan (48) design, to decrease
the dead volume.   The internal diameter of the bulb was decreased
from 20 mm to 15 mm, and that of the quartz atomizer was decreased
from 10 mm for a batch procedure to 7 mm, to increase the atomic
population in the light path.   These designs have been used successfully by the author's group in a number of applications (43,44,49).
     A microporous Teflon tube (3 mm ID, 4 mm OD, 70% porosity)
gas-liquid separator was used by Yamamoto for the determination

of arsenic in a flow injection hydride generation system (20). A one-fold better sensitivity was reported compared to results obtained with a U-tube separator. However, this has not been confirmed in our laboratory when a similar separator was substituted for the U-tube separator in the determination of arsenic and selenium using the manifold shown in Fig. 13. The microporous Teflon tube separator produced no improvement for arsenic whereas for selenium the sensitivity was even inferior to that using a U-tube separator. The difference in behavior might be due to differences in the dead volumes of the U-tube separators and properties of the different hydrides.

In the manifold used by Yamamoto (45) for the determination of bismuth, arsenic, antimony, selenium, and tellurium, nitrogen or air was introduced into the carrier stream before the injection port by a separate pump line to limit the sample dispersion. This was found to produce a one-fold enhancement in the sensitivity for arsenic compared to a system without air segmentation.

Flow injection hydride generation has also been used successfully with an ICP emission spectrometric system. Liversage et al. (47) used the system to determine arsenic, but the detection limit (8 μg L^{-1}) and precision (7.2% RSD) were inferior to those reported for atomic absorption systems.

7.2 Performance of Flow Injection Hydride Generation Systems

The performance data for typical flow injection hydride generation atomic absorption systems are shown in Table 1 together with data for typical conventional and air-segmented continuous flow hydride generation systems for the analysis of selenium and arsenic.

7.3 Tolerance of Interferences in Flow Injection Hydride Generation Systems

It has frequently been reported that flow injection hydride generation methods are more tolerant to interfering ions in the sample solution (6,43-46,49). Åström (42) has shown that in comparison to conventional batch methods, interference from cobalt, nickel, and copper, which comprise the most serious interfering elements is substantially suppressed in the flow injection system, as about 100 to 1000 times higher concentrations of interferents can be tolerated in the sample.

Wang and Fang (6,43) have reported 10 to 100-fold higher tolerable concentrations for most of the interferents studied compared to results obtained with a segmented continuous flow system of the Skeggs type in the determination of selenium. A similar phenomenon has been observed by Chan (46) in the determination of selenium in rocks with a flow injection hydride generation system.

TABLE 1   Comparison of Performance of Various Hydride Generation Atomic Absorption Spectrophotometric (HGAAS) Methods

Method	Detection limit[a] ($\mu g\ L^{-1}$)		Precision RSD (%)		Sample volume (mL)		Sampling frequencies (samples h^{-1})	
	As	Se	As	Se	As	Se	As	Se
Flow injection HGAAS (44, 49)	0.1 (0.04)	0.06 (0.02)	1.5	1.6	0.4	0.4	220	250
Batch HGAAS (50)	0.06 (0.6)	0.06 (0.6)	2.7	2.7	10	10	80	80
Segmented continuous flow HGAAS (51)	1 (4)	1 (4)	2	2	4	4	20	20

[a]Amounts in parentheses indicate detection limits in nanograms.

The reason for the high tolerance of interferences was often thought to be that the shorter residence time of the reaction mixture in flow injection systems is favorable for the formation of hydrides which are generally fast, while the slower interfering reactions are suppressed. Another reason might be that owing to the small sample volume injected in comparison to the sample consumption in a conventional batch method, a smaller absolute amount of the interfering species is introduced into the system, leaving less chance for their accumulation in the reactor and separator and hence less chance for their contact with the hydride. Existing theories and assumptions on the mechanisms of interference in hydride generation systems support the foregoing postulations. Interference from transition metals in hydride generation atomic absorption methods was often thought to be due to adsorption or decomposition of the hydrides by the finely dispersed metal precipitates formed during the reduction (52-54); but Bye (55) has proposed that the interference might be due to the formation of metal boride precipitates, which could decompose the hydrides before they reached the atomizer. In any case, the short residence time which is an important feature of all flow injection hydride generation systems would be beneficial for decreasing the probability of contact of the hydrides with reactive interfering species present in the system. Åström (42) and Wang and Fang (49) have shown that decreasing the residence time by using shorter reaction coils in a flow injection hydride generation system was beneficial for minimizing interferences in the determination of bismuth and arsenic, respectively. When 100 µg of copper per milliliter was present in a 25-µg L^{-1} bismuth standard, using a system similar to that in Fig. 13 but with mixing and reaction coil lengths a = 100, b = 100, c = 50 cm, the suppression of the bismuth signal was 100% (42), yet when the coil lengths were shortened so that a = 100, b = 0, c = 0 cm, even when the coexisting copper concentration was increased to 200 µg mL^{-1} no suppression of the bismuth has been observed.

It has been assumed that interferences in hydride generation atomic absorption spectrometric methods could also occur in the gas phase. According to Welz and Melcher (56), the mechanism of the gas phase interference might be competition of hydride-forming elements with active free radicals in the process of atomization. Meyer et al. (57) have shown that it is the *absolute* amount of the interferent which could react with the free radicals that determines the extent of interference, not the *relative* concentration of interferent to analyte. This could perhaps also explain why flow injection systems characterized by small sample consumption are more tolerant to interferents.

8   COLD VAPOR METHODS FOR DETERMINATION OF MERCURY

Cold vapor methods for the determination of mercury have been used extensively with atomic absorption spectrometry and to a smaller extent with atomic fluorescence spectrometry. The cold vapor technique has recently been adapted to flow injection systems with very promising results and interesting developments. Flow injection cold vapor systems are characterized by low sample consumption, high efficiency, and high sensitivity; the systems are easy to automate and are considerably simpler than most commercial models for cold vapor mercury determinations. These will be illustrated by the various manifolds discussed in the following sections.

8.1   Integrated Membrane Phase Separator-Absorption Cell System

The first flow injection system used for cold vapor determination of mercury by atomic absorption spectrometry was proposed by de Andrade (58). An ingenious flow cell was designed which combined a special vapor diffusion membrane separator with an absorption cell. The device was based on the permeability of commercial microporous Teflon tape to mercury vapor. On addition of reductant to the sample, which was injected into an acidic carrier stream, mercury vapor was generated in the stream and transported into a channel on one side of the gas-permeable Teflon membrane and diffused directly

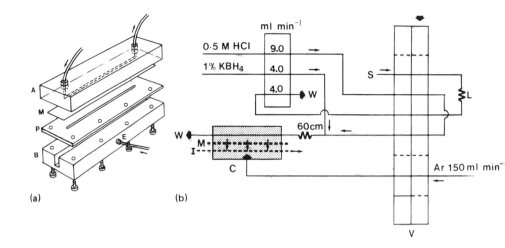

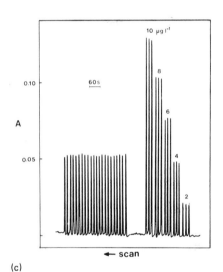

(a)

(b)

(c)

FIGURE 14   Integrated membrane phase separator-absorption cell
system for the cold vapor atomic absorption determination of mercury.
(a)   Detailed structure of the integrated phase separator-absorption
flow cell.   A, Plexiglass block with 1.5 × 110 mm channel at the bottom; B, plexiglass block with a 3 × 6 × 170 mm groove acting as absorption flow cell; M, Teflon membrane backed with nylon gauze on
the underside; P, 1-mm-thick plexiglass plate with 1.5 × 110 mm
slit in the center; E, entrance for argon.   (b)   Manifold of the system.   C, Integrated phase separator-absorption cell; L, sample loop,
400 μL; V, eight-channel multifunctional valve, dashed lines are
blocked channels; M, porous membrane; S, sample; I, incident light
from hollow-cathode lamp; W, waste.   (c)   Recordings for mercury
standards obtained with the system in (b).   Right, standard series
in the range 2 to 10 μg L^{-1} mercury; left, replicate injections of a
4-μg L^{-1} mercury standard.   [Reproduced with permission of Elsevier
Science Publishing (6).]

into the absorption cell, positioned in the light path of the spectrometer. The detection limit was 1.4 μg L^{-1} with a sampling frequency of 110 samples per hour.

Despite the simplicity and efficiency of the system, the sensitivity was lower than in many published systems for cold vapor
atomic absorption determinations of mercury. An attempt was made
in our laboratory to improve the detection limit and to prolong the
lifetime of the Teflon membrane (6,59). The de Andrade system
was modified by using a thinner microporous Teflon tape (0.075 mm
thick) and reinforced with a nylon gauze to increase its mechanical
strength. A longer and narrower diffusion channel, measuring
1.5 × 110 mm, compared to 3 × 50 mm for the previous design, but
having almost identical diffusion areas was used to limit the sample
dispersion. A 1-mm-thick plexiglass plate with a 1.5-mm slot along
the diffusion channel was inserted between the membrane and the
absorption cell to prevent the partial blockage of the light path
through fluctuations of the membrane when the carrier stream pul-
sated [Fig. 14(a)]. These measures increased the diffusion efficiency and lowered the baseline noise, producing large improvements
in the sensitivity and detection limits. A characteristic concentra-
tion of 1.3 μg L^{-1} and a detection limit of 0.06 μg L^{-1} with 400-μL
samples has been achieved. Sampling frequency was 200 samples
per hour and the precision was 1.0% RSD. Argon gas was directed
into the absorption cell after the peak response had been reached
by switching the multifunctional valve into sampling position [Fig.
14(b)]. This operation flushed the mercury vapor out of the cell
very rapidly and a high sampling frequency of 200 samples per hour
could be attained. The integrated membrane phase separator-absorption cell system has been further simplified and miniaturized without
reducing performance (60).

8.2 Cold Vapor Mercury Determinations Using Flow
Injection Hydride Generation Systems

Flow injection hydride generation systems such as those described
in Section 4.6 may also be used as cold vapor systems for the determination of mercury, with minor modifications in the flow parameters. The quartz atomizer does not need heating for such purposes.
The performance of such a system has been compared with the membrane phase separator system mentioned in the preceding section
and was found to be similar (6). It seems that better sensitivity
might be expected with the latter approach because of the absence
of a relatively large dead volume gas-liquid separator which is always included in flow injection hydride generation systems; however, probably because of incomplete transport of mercury through
the membrane, the sensitivity was not found to be superior. No

detailed information on the use of hydride generation systems for the cold vapor determination of mercury has hitherto been published, although it seems convenient to combine the two techniques by using the same hardware.

8.3 Cold Vapor Determination of Mercury by Flow Injection Atomic Fluorescence

Morita et al. (61) reported a flow injection system for the determination of mercury by atomic fluorescence. The sample was injected into an acid carrier and merged with the stannous chloride reductant solution downstream at a confluence point. After passing through a mixing coil the mercury vapor generated was separated from the liquid phase by passing the stream through a specially designed cylinder-type gas-liquid separator. A string of small glass beads was hung in a 12-mm-ID 200-mm-long glass cylinder in a vertical position. The reaction mixture was directed onto the bead string and the mercury vapor was evolved while the liquid flowed down the bead string and thence via an outlet tube to waste. Argon gas introduced at the cylinder base carried the mercury vapor through a trap to the atomic fluorescence detector. A detection limit of about 0.1 μg L^{-1} could be achieved with 64-μL sample injection, but the sampling frequency (35 samples per hour) is rather low compared to other flow injection cold vapor methods; this might be due to the long washout period resulting from the relatively large dead volume of the gas-liquid separator and trap.

9   METHODS FOR SEQUENTIAL AND SIMULTANEOUS DETERMINATIONS

Methods that could be used to perform the determination of two or more species in the same sample sequentially or simultaneously not only could improve efficiency but could also decrease sample consumption. The basic principles of the flow injection technique make it quite suitable for the development of such methods. This has been the topic of an excellent review by de Castro and Valcárcel (62).

In sequential determination systems several detectors are often connected in series; atomic spectroscopic detectors are quite suitable for this purpose. Samples processed in a preceding detector and subsequently introduced into an atomic spectroscopic detector are capable of producing analytical results without being influenced by any type of foregoing chemical reactions. However, despite the large potentials in such applications shown by recently published work, the publications to date are rather few and much remains to be explored.

Some atomic spectrometric instruments, such as the multichannel ICP spectrometer and atomic fluorescence spectrometer, are capable of simultaneous multicomponent determinations. In principle, the combination of flow injection with such detectors should be able to enhance the capabilities of both the spectrometric and flow injection techniques, although the publications are also few.

9.1 Sequential Methods for Speciation

Flame atomic absorption and ICP emission spectrometric detectors are often ideal for the determination of the total concentration of an analyte irrespective of its chemical form. The connection of such detection systems to other detectors in series (such as photometric or electrochemical detectors), which could provide specificity for a certain species (valence state, complexed form, etc.) of the analyte, could result in a very efficient speciation system which is capable of determining two or more species of the analyte in a single sample bolus. For obvious reasons the atomic spectrometric detector must always be placed last in a series of different detectors.

Lynch et al. (63) used a flow injection system with a spectrophotometer connected in series with an atomic absorption instrument for the determination of total chromium and chromium(VI), total iron, and iron(II). The diphenyl carbazide spectrophotometric method was used for the determination of chromium(VI) and the 1,10-phenanthroline method was used for iron(II), whereas the total metals were determined by atomic absorption spectrometry. A precision of 1 to 2% RSD has been achieved at a high sampling frequency of 120 samples per hour. A similar system for iron speciation has been described by Burguera and Burguera (69).

Speciation could also be achieved by using a single atomic spectrometric detector with sequential analysis involving a separation step. On-line ion-exchange columns could be used to retain one species while the other is determined by the detector. The one retained is then eluted into the same detector. Cox et al. (65) reported a flow injection ICP emission spectrometric system incorporating an activated alumina (acid form) minicolumn in the manifold for the determination of chromium(III) and chromium(VI) in synthetic solutions and natural waters. Pacey and Bubnis (66) and Milosavljevic et al. (67) have also used such systems for speciation studies. Silva et al. (68) used a flow injection system with atomic absorption detection and liquid-liquid extraction to achieve sequential indirect analysis of nitrite and nitrate. Methods involving separations are treated in more detail in Chapter 5.

9.2 Sequential Methods for Multielement Determinations

An atomic spectrometric detector could be connected in series with another flow-through detector in a flow injection manifold to achieve sequential determination of two or more elements, with the spectrometer placed last in the sequence of detectors. The capability of such a system is well demonstrated by Alonso and Bartroli (69) in their sequential system, which was composed of a tubular PVC membrane calcium electrode connected in series with an atomic absorption spectrometer. Calcium and magnesium in waters were determined sequentially when an injected sample passed through the two detectors. On-line dilution was included in the flow injection manifold to obtain linear response for the atomic absorption determination of magnesium. Good results were obtained for both elements at a sampling frequency of 60 to 90 samples per hour.

9.3 Simultaneous Multielement Determinations

Simultaneous multielement determinations could either be carried out by using several parallel instruments or by using a multichannel single instrument for detection. In the former approach the sample could be split up after injection and carried to different detectors. Basson and van Staden described a flow injection system for the simultaneous determination of four components (sodium, potassium, magnesium, and calcium) using a double-channel flame photometer and a double-channel atomic absorption spectrometer in parallel (23). The sample is injected into an aqueous carrier and merged with a reagent solution containing lanthanum, cesium, and lithium. The mixture is split up and fed into the two detectors, sodium and potassium being determined by flame photometry and calcium and magnesium by atomic absorption spectrometry. The sampling frequency was 128 samples per hour.

The multichannel ICP spectrometer has been used by several groups for simultaneous multielement analysis (12,14,15,70). The flow injection manifold is usually quite simple, involving the most elementary form of sample introduction, using an injection port or valve and a short connection line. The peak height readout mode could, of course, be used as in single-channel determinations, but obviously, multichannel recording with chart recorders will not be practical above a certain limit. As a result, integration of the peak area for each element is often used as the readout mode, the integration period beginning after a predetermined delay period following sample injection, synchronized with the rising of the peak and lasting for about 15 to 20 s. Alexander et al. (12) have also measured the peak heights by performing 3-s integrations at both the baseline and the peak maxima and subtracting the two values as a form of multichannel readout. In any case, as with most commercial spectrometers, synchronizing the injection device with the integration program might not be directly possible; coupling of a computer via an interface might be necessary.

A rapid and precise method for the simultaneous determination of sodium, potassium, calcium, magnesium, lithium, copper, iron, and zinc in blood serum by flow injection ICP emission spectrometry has been proposed by McLeod et al. (70) using aqueous synthetic multielement solutions for calibration. The 30-channel spectrometer was equipped with an Apple II computer for data acquisition. Satisfactory results for all elements studied were obtained using as little as 20 μL of sample. The peak area integration mode was used for quantification. Alexander et al. (12) determined 13 elements simultaneously in 10-μL samples with a multichannel ICP spectrometer. The precision for most elements was 1 to 3% RSD and a sampling frequency of 240 samples per hour could be achieved. The method was applied to serum analysis, and good agreement with values from a control serum was obtained even when using the least precise peak height approach for quantification. Internal standardization and peak area integration were both considered unnecessary in this application. Zagatto et al. used the generalized standard additions method of calibration (see Section 5.6) in the simultaneous determination of nickel, copper, and zinc in copper-nickel alloys working with a flow injection multichannel ICP spectrometric system (31). A 10-s integration period was used to evaluate the peak areas.

9.4 Gradient Scanning Simultaneous Multielement Determinations

The gradient scanning technique involves rapid scanning of two physical parameters (e.g., wavelength versus absorbance or intensity) at a certain point on the concentration gradient of a well-defined dispersed sample zone (71). This technique was applied by Fang et al. (32) to perform simultaneous determinations of sodium, potassium, lithium, and calcium by flame photometry, using a fast-scanning monochromator capable of scanning from 350 to 800 nm in 5 ms. A storage oscilloscope was used to store the scanning data, which were later retrieved by plotting on an X-Y recorder. The three-dimensional presentation in Fig. 15 shows successive scannings of a tap-water sample at 0.3-s intervals. The intensity ranges for each of the analytes could be optimized separately by scanning on different sections of the peak gradient during a single injection of the sample. Thus for an element with high emission intensity, a scanning on the tailing part of the gradient might improve the linearity of the calibration curve, whereas a scanning on the peak maximum will be preferred for an element with low intensity. With

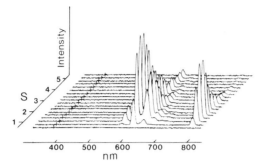

FIGURE 15   Three-dimensional gradient-scanning recording of a
40-μL sample of tap water.   Time between successive scannings
0.3 s.   Wavelengths of peaks, left to right: CaOH, 554 nm; Na,
589 nm; CaOH, 622 nm; K, 767 nm.   [Reproduced with permission
of the American Chemical Society (32).]

adequate computer facilities, multielement standard addition calibration with standard/sample ratios optimized for each individual element should also be possible with the gradient scanning technique,
as shown in Section 5.7.

Gradient exploitation should also be possible with a multichannel ICP spectrometer by using short integration periods carried out
at predetermined delay times after sample injection; however, most
commercial instruments might not be able to perform and store more
than one integration and have the data processed during each cycle
of injection, so that considerable adaptation work might be necessary.
The scanning monochromator mentioned above had very low resolution, with a spectral bandwidth of 10 nm, and hence could not be
used with emission sources with more complex spectra, other than
that of a low-temperature flame.   Ultrafast-scanning emission spectrometers with adequate resolution and appropriate fast data collection and storage facilities will be needed to explore fully the potentials of this technique.

10   OTHER METHODS AND TECHNIQUES

10.1   Methods for Investigation of Interference Effects

The study of interferences in atomic spectroscopy always involves
the addition of various reagents containing the interferents to the
sample solutions.   If different concentration ratios for analytes and
interferents were to be studied for a large range of interferents,
the investigation will involve considerable volumetric manipulation.
It has been shown in the preceding sections that flow injection

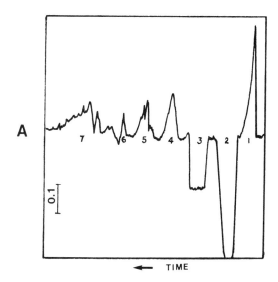

FIGURE 16 Study of interference effects in atomic absorption spec-troscopy using manifold in Fig. 2. X, Calcium stream; C, water; interferent solution injected at S. 1, 100 µg mL^{-1} calcium; 2, 1000 ppm aluminum; 3, 1000 ppm phosphate; 4, ethanol; 5, MIBK; 6, 1000 ppm potassium; 7, 10% lanthanum. [Reproduced with permis-sion of the Royal Society of Chemistry (19).]

technique may be a powerful tool in simplifying solution handling in analytical processes; the same applies to method development work such as the study of interference effects. This capability has been illustrated by Tyson et al. (19) using a manifold identical to that used for dilution in Fig. 2 to study interference effects in atomic absorption spectrometry. The analyte solution under study was pumped through the diluent line and the interferent was injected into a water carrier in a second line which was merged with the analyte solution downstream. Concentration of the interferent in the final solution could easily be computed by evaluating the dis-persion coefficient of the manifold.

The interference effects of aluminum, phosphate, ethanol, MIBK, potassium, and lanthanum on the absorbance of calcium were studied by observing the positive or negative peaks produced on the base-line of the analyte response. The recordings from this study are shown in Fig. 16. It is quite interesting to note that the profiles of the interfering peak of different interferents are quite varied. The depressive effects of phosphate and aluminum are obvious, whereas lanthanum, potassium, and the organic solvents show different degrees of enhancement. The 100 µg mL^{-1} calcium peak profile reflects the actual concentration gradient of the injected material. The plateau on the phosphate peak implies a critical concentration for the interference of phosphate below which the depression effect on calcium increases with increasing concentration of phosphate, but remains constant above the critical concentration. The behavior of the interference from aluminum is quite different in that the depression increases continuously with increase in aluminum concentration until the response from calcium disappears completely. It could be seen from the above that the technique could provide rapid assessment of interference effects at different concentration levels and also information on variations with changes in operation parameters.

10.2 Flow Injection Technique as an Interface Between HPLC and Atomic Absorption Spectrometer

High-performance liquid chromatography (HPLC) is a powerful tool for the separation of different species in the liquid phase. When atomic absorption is used as a detector, HPLC could be used for speciation studies of metals or for the elimination of interferences. Although the combination of the two techniques implies combination of high separation efficiency with a high specificity for detection, which should result in a very powerful technique, the coupling is not without problems. The elution rate of HPLC is normally much smaller than the normal aspiration rate of an atomic absorption spectrometer and the performance of one or both of the instruments will have to be compromised in the coupling.

The flow injection technique has been used by Renoe at al. (72) to provide an interface for the coupling of HPLC to atomic absorption instruments. The interface, called the "flow injection sample manipulator," allows optimization of performance for both the HPLC and atomic absorption instruments. A small volume of the HPLC eluate is sampled with a flow injection valve during the elution at the optimal elution rate and is then presented to the atomic absorption spectrometer by injection into a carrier stream. On-line dilution and reagent addition is easily performed after injection, resulting in a final total flow rate compatible with that of the atomic absorption instrument. Successive injections at regular intervals during the elution produce a recording of a series of flow injection peaks, the peak maxima of which describe the profile of the HPLC elution process in relation to the metal under consideration. Renoe used the system successfully for the separation and determination of citrate-bound protein associated and free forms of calcium and magnesium. The recording for the determination of citrate-bound and free calcium is shown in Fig. 17. The interface was found to

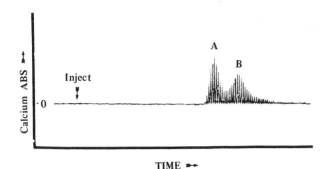

FIGURE 17   HPLC chromatograms from an aqueous sample containing calcium and citrate, obtained by using flow injection sample manipulator as an interface between HPLC and AAS.   Peak A is the calcium citrate complex and peak B is noncomplexed calcium. [Reproduced with permission of the American Chemical Society (72).]

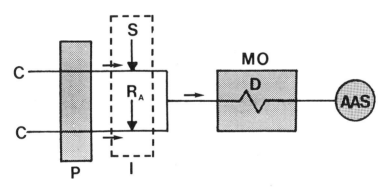

FIGURE 18   On-line microwave-oven sample decomposition flow injection system for the determination of copper, zinc, and iron in whole blood.   C, Carrier; P, pump; I, double-channel injector; D, decomposition tube; MO, microwave oven; AAS, atomic absorption spectrometer.   Sample and acid reagent are introduced into carrier at S and R_A.   [Reproduced with permission of Elsevier Science Publishing (73).]

be capable of achieving virtually simultaneous separation by HPLC and detection by atomic absorption spectroscopy without compromising the performance of either instrument.

10.3 On-Line Sample Decomposition Using Microwave Oven

Burguera et al. (73) have proposed a simple and rapid method for the determination of copper, zinc, and iron in whole blood with on-line sample decomposition. The setup is shown in Fig. 18. Sixty to one hundred microliters of whole blood was injected and merged with an acid reagent. The mixture was passed through a Pyrex coiled decomposition tube fixed in a commercial domestic microwave oven. The mineralized sample flowed directly into the atomic absorption instrument and the metals were determined. The optimum residence time for the decomposition was found to be within 15 to 26 s, making possible a high sampling frequency of 80 samples per hour. Recoveries were good and the relative standard deviation was less than 3%. Besides being efficient and reliable, the technique also minimized problems in contamination and generation of acid fumes during digestion.

REFERENCES

1, R, F, Browner and A. W. Boorn, Sample introduction: The Achilles' heel of atomic spectroscopy? *Anal. Chem.*, *56*:786A (1984).

2. A. J. Faske, K. R. Snable, A. W. Boorn, and R. F. Browner, Microliter sample introduction for ICP-AES, *Appl. Spectrosc.*, *39*:542 (1985).

3. W. R. Wolf and K. K. Stewart, Automated multiple flow injection analysis for flame atomic absorption spectrometry, *Anal. Chem.*, *51*:1201 (1979).

4. B. F. Rocks, R. A. Sherwood, and C. Riley, Direct determination of therapeutic concentrations of lithium in serum by flow injection analysis with atomic absorption spectroscopic detection, *Clin. Chem.*, *28*:440 (1982).

5. N. Zhou, W. Frech, and E. Lundberg, Rapid determination of lead, bismuth, antimony and silver in steels by flame atomic absorption spectrometry combined with flow injection analysis, *Anal. Chim. Acta*, *153*:23 (1983).

6. Z. Fang, S. Xu, X. Wang, and S. Zhang, Combination of flow injection techniques with atomic spectrometry in agricultural and environmental analysis, *Anal. Chim. Acta*, *179*:325 (1986).

7. J. M. Harnly and G. R. Beecher, Signal to noise ratios for flow injection atomic absorption spectrometry, *J. Anal. At. Spectrom.*, *1*:75 (1986).

8. K. Fukamachi and N. Ishibashi, Flow injection atomic spectrometry with organic solvents, *Anal. Chim. Acta*, *119*:383 (1980).

9. A. S. Attiyat and G. D. Christian, Nonaqueous solvents as carrier or sample solvent in flow injection analysis/atomic absorption spectrometry, *Anal. Chem.*, *56*:439 (1984).

10. T. Ito, H. Kawaguchi, and A. Mizuike, Inductively coupled plasma emission spectrometry of microliter samples by a flow injection technique (in Japanese), *Bunseki Kagaku*, *29*:334 (1980).

11. S. Zhang, L. Sun, H. Jiang, and Z. Fang, Determination of copper, zinc, iron, manganese, sodium, potassium, calcium, and magnesium in plants and soil by flow injection atomic absorption spectrometry (in Chinese), *Guangpuxue yu Guangpu Fenxi*, *4*(3):42 (1984).

12. P. W. Alexander, R. J. Finlayson, L. E. Smythe, and A. Thalib, Rapid flow analysis with inductively-coupled plasma atomic emission spectroscopy using a micro-injection technique, *Analyst*, *107*:1335 (1982).

13. M. W. Brown and J. Růžička, Parameters affecting sensitivity and precision in the combination of flow injection analysis with atomic absorption spectrophotometry, *Analyst*, *109*:1091 (1984).

14. A. O. Jacintho, E. A. G. Zagatto, H. Bergamin F⁰., F. J. Krug, B. F. Reis, R. E. Bruns, and B. R. Kowalski, Flow injection systems with inductively-coupled argon plasma atomic emission spectrometry. Part 1. Fundamental considerations, *Anal. Chim. Acta*, *130*:243 (1981).

15. S. Greenfield, FIA weds ICP—a marriage of convenience, *Ind. Res. Dev.*, *23*:140 (1981).

16. N. Yoza, Y. Aoyagi, S. Ohashi, and A. Tateda, Flow injection system for atomic absorption spectrometry, *Anal. Chim. Acta*, *111*:163 (1979).

17. B. F. Rocks, R. A. Sherwood, and C. Riley, Direct determination of calcium and magnesium in serum using flow injection analysis and atomic absorption spectroscopy, *Ann. Clin. Biochem.*, *21*:51 (1984).

18. B. D. Mindel and B. Karlberg, A sample pretreatment system for atomic absorption using flow injection analysis, *Lab. Pract.*, *30*:719 (1981).

19. J. F. Tyson, C. E. Adeeyinwo, J. M. H. Appleton, S. R. Bysouth, A. B. Idris, and L. L. Sarkissian, Flow injection techniques of method development for flame atomic absorption spectrometry, *Analyst*, *110*:487 (1985).

20. M. Yamamoto, M. Yasuda, and Y. Yamamoto, Flow injection-hydride generation-atomic absorption spectrometry with gas diffusion unit using microporous PTFE membrane (in Japanese), *J. Flow Inject. Anal.*, *2:*134 (1985).

21. E. A. G. Zagatto, F. J. Krug, H. Bergamin F⁰., S. S. Jørgensen, and B. F. Reis, Merging zones in FIA. Part 2. Determination of calcium, magnesium and potassium in plane material by continuous flow injection atomic absorption and flame emission spectrometry, *Anal. Chim. Acta, 104:*279 (1979).

22. R. F. Reis, A. O. Jacintho, J. Mortatti, F. J. Krug, E. A. G. Zagatto, H. Bergamin F⁰., and L. C. R. Pessenda, Zone sampling processes in flow injection analysis, *Anal. Chim. Acta, 123:*221 (1981).

23. W. D. Basson and J. F. van Staden, Simultaneous determination of sodium, potassium, magnesium and calcium in surface, ground and domestic water by flow injection analysis, *Fresenius Z. Anal. Chem., 302:*370 (1980).

24. J. F. Tyson, Extended calibration of flame atomic absorption instruments by a flow injection peak width method, *Analyst, 109:*319 (1984).

25. J. Tyson, Flow injection techniques for flame atomic absorption spectrophotometry, *Trends Anal. Chem., 4:*124 (1985).

26. J. F. Tyson, Low cost continuous flow analysis. Flow injection techniques in atomic absorption spectrometry, *Anal. Proc. (London), 18:* 542 (1981).

27. S. Greenfield, Inductively coupled plasma-atomic emission spectroscopy with flow injection analysis, *Spectrochim. Acta, Part B, 38:* 93 (1983).

28. J. F. Tyson, J. M. H. Appleton, and A. B. Idris, Flow injection calibration methods for atomic absorption spectrometry, *Anal. Chim. Acta, 145:*159 (1983).

29. Y. Israel and R. M. Barnes, Standard addition method in flow injection analysis with inductively coupled plasma emission spectrometry, *Anal. Chem., 56:*1188 (1984).

30. B. E. H. Saxberg and B. R. Kowalski, Generalized standard addition method, *Anal. Chem., 51:*1031 (1979).

31. E. A. G. Zagatto, A. O. Jacintho, F. J. Krug, B. F. Reis, R. E. Bruns, and M. C. U. Araújo, Flow injection systems with inductively-coupled argon plasma atomic emission spectrometry. Part 2. The generalized standard addition method, *Anal. Chim. Acta, 145:*169 (1983).

32. Z. Fang, J. M. Harris, J. Růžička, and E. H. Hansen, Simultaneous flame photometric determination of lithium, sodium, potassium, and calcium by flow injection analysis with gradient scanning standard addition, *Anal. Chem., 57:*1457 (1985).

33. M. C. U. Araújo, C. Pasquini, R. E. Bruns, and E. A. G. Zagatto, A fast procedure for standard addtions in flow injection analysis, *Anal. Chim. Acta, 171:*337 (1985).

34. E. H. Hansen, J. Růžička, F. J. Krug, and E. A. G. Zagatto, Selectivity in flow injection analysis, *Anal. Chim. Acta, 148:* 111 (1983).

35. E. A. G. Zagatto, M. F. Giné, E. A. N. Fernandes, B. F. Reis, and F. J. Krug, Sequential injections in flow systems as an alternative to gradient exploitation, *Anal. Chim. Acta, 173:*289 (1985).

36. J. M. Ottaway, D. T. Coker, and J. A. Davies, Sensitive indirect method for the determination of titanium by AAS using an air-acetylene flame, *Anal. Lett., 3:*385 (1970).

37. P. Martínez-Jiménez, M. Gallego, and M. Valcárcel, Indirect atomic absorption determination of aluminium by flow injection analysis, *Microchem. J., 34:*190 (1986).

38. P. Martínez-Jiménez, M. Gallego, and M. Valcárcel, Indirect atomic absorption determination of cerium and lanthanum by flow injection analysis using an air-acetylene flame, *At. Spectrosc., 6:*137 (1985).

39. P. Martínez-Jiménez, M. Gallego, and M. Valcárcel, Indirect atomic absorption determination of uranium by flow injection analysis using an air-acetylene flame, *At. Spectrosc., 6:*65 (1985).

40. A. T. Haj-Hussein, G. D. Christian, and J. Růžička, Determination of cyanide by atomic absorption using a flow injection conversion method, *Anal. Chem., 58:*38 (1986).

41. B. A. Petersson, Z. Fang, J. Růžička, and E. H. Hansen, Conversion techniques in flow injection analysis. Determination of sulfide by precipitation with cadmium ions and detection by atomic absorption spectrometry, *Anal. Chim. Acta, 184:*165 (1986).

42. O. Åström, Flow injection analysis for the determination of bismuth by atomic absorption spectrometry with hydride generation, *Anal. Chem., 54:*190 (1982).

43. X. Wang and Z. Fang, Determination of trace amounts of selenium in environmental samples by hydride generation atomic absorption spectrometry combined with flow injection analysis, *Fenxi Huaxue, 14:*738 (1986) (in Chinese).

44. X. Wang and Z. Fang, Determination of trace amounts of selenium in environmental samples by hydride generation-atomic absorption spectrometry combined with flow injection analysis technique, *Kexue Tonghao (Science Bulletin), 31:*791 (1986).

45. M. Yamamoto, M. Yasuda, and Y. Yamamoto, Hydride generation atomic absorption spectrometry coupled with flow injection analysis, *Anal. Chem., 57:*1382 (1985).

46. C. Y. Chan, Semiautomated method for the determination of selenium in geological materials using a flow injection analysis technique, *Anal. Chem.*, *57*:1482 (1985).

47. R. R. Liversage, J. C. van Loon, and J. C. de Andrade, A flow injection/hydride generation system for the determination of arsenic by inductively-coupled plasma atomic emission spectrometry, *Anal. Chim. Acta*, *161*:275 (1984).

48. P. N. Vijan and G. R. Wood, An automated submicrogram determination of arsenic in atmospheric particulate matter by flameless atomic absorption spectrophotometry, *At. Absorpt. Newsl.*, *13*:33 (1974).

49. X. Wang and Z. Fang, The determination of arsenic in environmental samples by flow injection hydride generation atomic absorption spectrometry (in Chinese), *Fenxi Huaxue*, *16*:912 (1988).

50. Z. Fang, S. Zhang, and H. Jiang, Hydride generation atomic absorption determination of trace amounts of arsenic, and selenium in waters, soils, grains and biological materials (in Chinese), *Bull. Inst. For. Pedol. Acad. Sin.*, *4*:139 (1980).

51. K. S. Subramanian, Rapid electrothermal atomic absorption method for arsenic and selenium in geological material via hydride evolution, *Fresenius Z. Anal. Chem.*, *305*:382 (1981).

52. A. E. Smith, Interferences in determination of elements that form volatile hydrides with sodium borohydride using AAS and the argon-hydrogen flame, *Analyst*, *100*:300 (1975).

53. B. Welz and M. Melcher, Mechanisms of transition metal interferences in hydride generation AAS. Part 1. Influence of Co, Cu, Fe and Ni on Se determination, *Analyst*, *109*:569 (1984).

54. B. Welz and M. Schubert-Jacobs, Mechanisms of transition metal interferences in hydride generation atomic absorption spectrometry. Part 4. Influence of acid and tetrahydroborate concentrations on interferences in arsenic and selenium determinations, *J. Anal. At. Spectrom.*, *1*:23 (1986).

55. R. Bye, Interferences from bivalent cations in the determination of selenium by hydride-generation and atomic-absorption spectrometry, *Talanta*, *33*:705 (1986).

56. B. Welz and M. Melcher, Mutual interactions of elements in the hydride techniques in atomic absorption spectrometry. Part 1. Influence of selenium on arsenic determination. *Anal. Chim. Acta*, *131*:17 (1981).

57. A. Meyer, Ch. Hofer, G. Tolg, S. Raptis, and G. Knapp, Cross-interferences by elements in the determination of traces of selenium by the hydride-atomic absorption spectrometric procedure (in German), *Fresenius Z. Anal. Chem.*, *296*:337 (1979).

58. J. C. de Andrade, C. Pasquini, W. Baccan, and J. C. van Loon, Cold vapor atomic absorption determination of mercury by flow injection analysis using a Teflon membrane phase separator coupled to the absorption cell, *Spectrochim. Acta, Part B, 38*:1329 (1983).

59. S. Zhang, Z. Fang, and J. Sun, Cold vapor atomic absorption determination of mercury in soils and plants using a flow injection gas diffusion system (in Chinese), *Guangpuxue yu Guangpu Fenxi, 6*:31 (1986).

60. S. Zhang, S. Xu, and Z. Fang, Determination of mercury by cold vapor AAS using on-line ion-exchange preconcentration with an integrated membrane phase separator absorption cell system, *Fenxi Huaxue*, in press (in Chinese).

61. H. Morita, T. Kimoto, and S. Shimomura, Flow injection analysis of mercury/cold vapor atomic fluorescence spectrophotometry, *Anal. Lett., 16*:1187 (1983).

62. M. D. Luque de Castro and M. Valcárcel, Simultaneous determinations in flow injection analysis, *Analyst, 109*: 413 (1984).

63. T. P. Lynch, N. J. Kernoghan, and J. N. Wilson, Speciation of metals in solution by flow injection analysis. Part 1. Sequential spectrophotometric and atomic-absorption detectors, *Analyst, 109*:839 (1984).

64. J. L. Burguera and M. Burguera, Flow injection spectrophotometry followed by atomic absorption spectrometry for the determination of iron(II) and total iron, *Anal. Chim. Acta, 161:* 375 (1984).

65. A. G. Cox, I. G. Cook, and C. W. McLeod, Rapid sequential determination of chromium(III)-chromium(VI) by flow injection analysis-inductively coupled plasma atomic emission spectrometry, *Analyst, 110:* 331 (1985).

66. G. E. Pacey and B. P. Bubnis, Flow injection analysis as a tool for metal speciation, *Am. Lab., 16:*17 (1984).

67. E. B. Milosavljevic, J. Růžička, and E. H. Hansen, Simultaneous determination of free and EDTA-complexed copper ions by flame atomic absorption spectrometry with an ion-exchange flow injection system, *Anal. Chim. Acta, 169*:321 (1985).

68. M. Silva, M. Gallego, and M. Valcárcel, Sequential atomic absorption spectrometric determination of nitrate and nitrite in meats by liquid-liquid extraction in a flow injection system, *Anal. Chim. Acta, 179*:341 (1986).

69. J. Alonso and J. Bartroli, Sequential flow injection determinations of calcium and magnesium in waters, *Anal. Chim. Acta, 179*:503 (1986).

70. C. W. McLeod, P. J. Worsfold, and A. G. Cox, Simultaneous multi-element analysis of blood serum by flow injection-inductively coupled plasma atomic emission spectrometry, *Analyst*, *109*:327 (1984).

71. J. Růžička and E. H. Hansen, Recent developments in flow injection analysis: Gradient techniques and hydrodynamic injection, *Anal. Chim. Acta*, *145*:1 (1983).

72. B. W. Renoe, C. E. Shideler, and J. Savory, Use of a flow injection sample manipulator as an interface between a "high-performance" liquid chromatograph and an atomic absorption spectrophotometer, *Clin. Chem.*, *27*:1546 (1981).

73. M. Burguera, J. L. Burguera, and O. M. Alarcón, Flow injection and microwave-oven sample decomposition for determination of copper, zinc and iron in whole blood by atomic absorption spectrometry, *Anal. Chim. Acta*, *179*:351 (1986).

5
Separation Techniques

MIGUEL VALCÁRCEL and MERCEDES GALLEGO *Department of Analytical Chemistry, University of Córdoba, Córdoba, Spain*

1 INTRODUCTION

Notwithstanding the fact that papers dealing with unsegmented continuous flow configurations involving analytical separation techniques
amount to only 10% of the total number of papers devoted to such
configurations (1800), they have shown a steadily growing trend in
the past few years (1) and will foreseeably account for a much
larger proportion by the end of this decade.

This chapter is concerned both with continuous systems where
the sample is injected [flow injection analysis (FIA) methods] and
with those involving its aspiration without air bubbles (unsegmented
flow methods). Their similarities justify their frequent joint treatment.

An FIA system allows for incorporation of separation systems
involving a variety of interfaces (2,3): gas-liquid (gas diffusion,
distillation, hydride generation), solid-liquid (precipitation, ion exchange, extraction), and liquid-liquid (dialysis, extraction). The
greater development of this association in FIA compared to air-segmented methodologies is probably the result of overcoming a series
of technical problems arising from the passage of the segmented
flow through the continuous separation system which normally result in perturbations decreasing the reproducibility achieved in the
measurements. This is not always so, as, for example, dialyzers
are used successfully with both types of continuous configuration.

The second phase used in continuous separation systems incorporated into FIA setups can either be introduced in a continuous way (e.g., hydride generation, dialysis, liquid extraction), be
generated in situ (precipitation), or be a permanent part of the
separation system (ion exchange).

The continuous separation process can involve a chemical
change (e.g., the formation of a gas, a chelate, or an ion pair),
a physical change (e.g., a temperature rise), or both, as a means
of facilitating the transfer of matter across the interface. This
can be rather small (e.g., liquid-liquid extraction, gas diffusion)
or large (precipitation, ion exchange). Dialysis, on the other hand,
requires no chemical or physical change to take place in the system.

The use of atomic spectroscopic detectors in conjunction with
FIA configurations involving on-line separation techniques results
in a number of advantages of particular interest. On the one hand,
they result in decreased human participation and reagent consumption and increased sample throughput as compared to conventional
analytical procedures. On the other hand, they provide higher
sensitivity and selectivity than do atomic spectroscopic methods re-
lying on direct introduction of sample, as a result of preconcentration and interference removal (matrix effects), respectively. Inso-
far as the preconcentration factor is a function of the sample volume

handled, the sample should preferably be introduced into the system by aspiration at a fixed flow rate over a given time interval. The increased selectivity achieved is a result of the differential separation of the sample components based on their uneven distribution in the phase according to their thermodynamic properties, aided in this case by kinetic discrimination (4). The interval during which the phases are in contact in these assemblies is somewhat short compared with that typical of conventional separation systems.

In addition to the advantages named above, the association of FIA with continuous separation techniques results in some others of undeniable interest. It allows one to carry out indirect atomic spectroscopic determinations on the basis of a prior reaction between the analyte and a metal cation acting as reagent, taking place after or simultaneously with the separation process. On-line configurations allow for expansion of the scope of application of atomic spectroscopic techniques otherwise tedious to implement by conventional procedures (5).

In this chapter we present and discuss the essential philosophical and practical aspects and the latest trends in continuous separations coupled on-line with absorption (AAS) and emission (ICP-AES) atomic spectroscopic instruments. The separation techniques considered here involve solid-liquid (precipitation, ion exchange) or liquid-liquid (extraction) interfaces. FIA systems involving gas-liquid interfaces (hydride generation) are dealt with separately in Chapter 4.

2 CONTINUOUS PRECIPITATION

Precipitation is a separation technique of wide use in classical analytical chemistry (6). Yet it has scarcely been used in automated procedures because of its inherent shortcomings, which are worsened if any intermediate precipitate removal step is involved. Some filtration devices of industrial application (rotating drums, plates, conveyor belt feeds) are useful to remove suspended substances prior to introduction of the fluid into the analyzer.

There are few automatic methods based on precipitate formation in systems other than turbidimetric or nephelometric (3,7-9). Skinner and Docherty (10) reported an air-segmented flow system for determination of potassium in fertilizers based on the removal of the tetraphenyl borate precipitate formed by means of a Technicon device consisting of a circle of filter paper moving perpendicularly to the flow. This system has not gained widespread acceptance probably because the results yielded are not as reproducible as those provided by other continuous separation systems, such as liquid-liquid extraction.

The use of precipitation as a separation technique in AAS has two basic aims: (a) the preconcentration of metal traces, generally by coprecipitation (11,12), (b) the development of indirect determinations for species precipitating with a given cation reagent (5), and (c) interference removal.

Our team recently developed the first flow injection configurations involving continuous precipitation, filtration, and dissolution (13). Růžička et al. (14) have devised a flow injection precipitation system with no filtration or dissolution and a narrower scope of application.

2.1 Basic Components of a Continuous Precipitation Device

The usual operation of a continuous precipitation system involves introducing the sample containing the analyte into a carrier solution including the precipitating reagent—usually a cation—the mixing of both resulting in the formation of a precipitate that is retained on a suitable filter. This is located prior to the measuring instrument and allows passage of the filtrate, which proceeds on to the detector. Thus the analyte is determined by the difference between the initial concentration of the precipitating reagent in the carrier and in the filtrate (nondissolution method). Alternatively, the filtrate can be wasted and the analyte determined after dissolution (precipitate dissolution method). It is thus apparent that a continuous precipitation system will involve two essential elements, the precipitation coil and the filter.

The *precipitation coil*, usually a helically coiled piece of Teflon tubing 0.3 to 0.7 mm in inner diameter and between 100 and 300 cm in length, ensures 100% precipitation provided that a suitable residence time for the analyte and the precipitating species is chosen.

The *filter* is used to quantitatively separate the precipitate from the supernatant. The filters used in continuous separation techniques implemented in FIA systems are of two basic types and were originally meant for use as cleaning devices in high-performance liquid chromatography (HPLC). One such filter is basically a column provided with a removable screen-type stainless steel filter of 0.5 μm pore diameter, 580 μL inner volume, and roughly 3 cm^2 filtration area. The other filter is also made of stainless steel and consists of a column of variable inner volume and several filters 0.5 to 2 μm in pore diameter with a filtration area of 7 mm^2. Both types of filters are depicted in Fig. 1. Experiments carried out in our laboratory have shown that

1. No significant changes in the signal result from the use of columns of 500 to 800 μL inner volume with filters of 0.5 to 2 μm

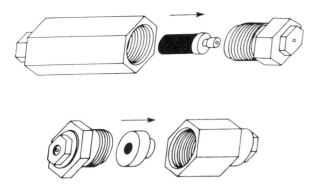

FIGURE 1 Continuous filtration devices: cylindrical and planar.

pore size. Large pore sizes (300 μm), though, give highly irreproducible results arising from passage of the precipitate through them.

2. Small chamber inner volumes (100 μL) call for filters with pore diameters greater than 1 μm.

3. For a given column and filter geometry, increasing amounts of precipitate result in head losses through friction. This, according to the Bernouilli equation, shows in a decrease in the flow rate of the stream emerging from the column with increase in the head loss. The effect is more marked when low-inner volume columns or small filtration areas are used.

4. The smaller the chamber inner volume or filtration area is, the more often the column and the filter should be cleaned.

2.2 Types of Configurations

As stated above, analytical determinations in continuous precipitation systems can be carried out either in the filtrate or from the precipitate—once dissolved. Thus there will be two principal types of configuration, discussed below.

2.2.1 Without Precipitate Dissolution

This, in turn, has two variants according to whether the normal (nFIA) or reversed (rFIA) flow injection mode is applied. In nFIA configurations [Fig. 2(a)], the sample containing the analyte (anion) is injected into a carrier stream containing excess precipitating cation. Once the reaction has completed, the precipitate formed is retained in the filter. The signal yielded by the cation, measured continuously at the detector, decreases proportionally to the concentration of the injected sample as the precipitate forms. The

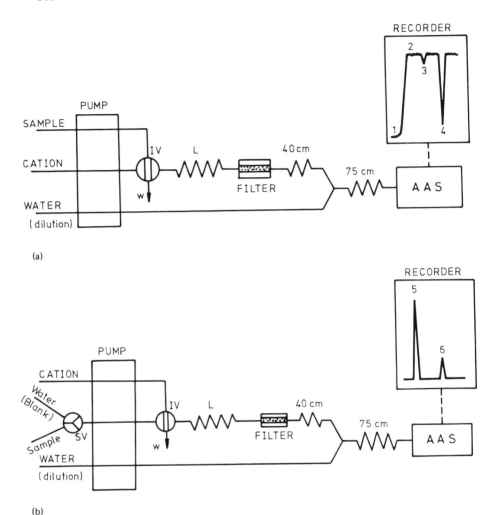

FIGURE 2 Precipitation manifold without dissolution for indirect
determinations of anions: (a) normal FIA; (b) reversed FIA. Recordings corresponding to: 1, baseline; 2, cation; 3, water injection; 4, sample injection; 5, cation injection into water carrier;
6, cation injection into sample carrier. L, precipitation coil; IV
and SV, injection and selecting valves, respectively. [Reproduced
with permission of the Royal Society of Chemistry (16).]

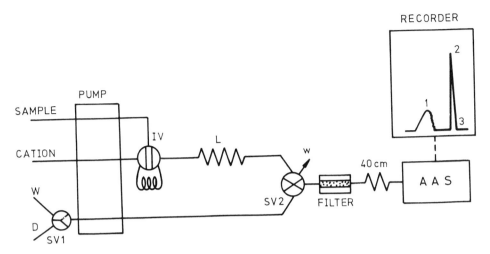

FIGURE 3 Precipitation manifold with dissolution for indirect determinations of anions. W, washing solution; D, dissolving solution.
Recordings: 1 and 2, signals corresponding to the precipitate washing and dissolution, respectively; 3, baseline; IV and SV, injection
and selecting valves, respectively; L, precipitation coil. [Repro-
duced with permission of the American Chemical Society (15).]

system includes a water stream for dilution of the carrier solution
prior to entering the nebulizer and uses water as a blank.

In rFIA setups [Fig. 2(b)], the reagent (precipitating cation)
is inserted into a carrier containing the analyte. Prior to the determination proper, the cation is injected into a water stream used
as blank and a positive peak is recorded for reference. Then the
selecting valve is switched to inject the cation in a stream containing the analyte, and a positive, smaller peak is obtained as a result of the precipitate formation. The difference between both
peaks is proportional to the analyte concentration in the sample.

2.2.2 With Precipitate Dissolution

This type of configuration (Fig. 3) includes a valve located prior
to the filter. Once the precipitate has been retained, the valve
allows the filter to be washed and then dissolved by the use of
appropriate solutions. Once the precipitate has been dissolved,
the cation concentration, proportional to that of the analyte in the
sample, is duly measured. This type of configuration has been
used solely with nFIA, as it yields poor results when it is the reagent that is injected into the stream containing the analyte (rFIA).

2.3   Theoretical and Practical Considerations

2.3.1   Types of Precipitate

Precipitates of analytical interest have very different physical properties that determine both their analytical application and the optimum experimental conditions for their formation.   They can be classified as crystalline (relatively pure and ready to filter), curdy (consisting of colloid particles forming larger aggregates of filterable size), and gelatinous (also colloid in nature, but more difficult to filter than curdy precipitates).

By the joint use of a continuous precipitation–filtration system and FIA, our team (15,16) has carried out a comprehensive study on the three aforesaid types of precipitate: crystalline (calcium oxalate), curdy (silver chloride), and gelatinous (ferric hydroxide).

2.3.2   Adsorption

Precipitate contamination is the principal source of error in gravimetric analysis.   The process via which impurities are incorporated into precipitates can take place both at the surface or within the precipitate.   Superficial impurification of precipitates may result from adsorption or postprecipitation, while inner impurification may arise from occlusion or coprecipitation.   Adsorption resulting from passage of the carrier solution containing the cation (nFIA) through the precipitate is the commonest cause of impurification in continuous precipitation systems.   The primary layer will preferentially absorb the excess ion common to the precipitate (i.e., the precipitating cation).   This type of impurity yields anomalous results, as it is the very species to be measured by AAS.   Of the three types of precipitate named above, curdy and gelatinous tend to adsorb the cation on their surface to a greater extent than does the crystalline type. The problem arising from the adsorption of cations on the precipitate is overcome by using an FIA system involving a washing step prior to the precipitate dissolution (15).   Reversed FIA systems, where the carrier is never the cation, are free from risk of adsorption of cations on the precipitate (16a).

2.3.3   Linkage of the Nebulizer Feed System and the Precipitation System

Continuous precipitation systems entail the use of large excesses of the precipitating cation to ensure high precipitation yields.   Thus, when inorganic precipitates are involved, the precipitating cation is used at concentrations between 25 and 80 $\mu g\ mL^{-1}$ (15).   Even greater concentrations (above 200 $\mu g\ mL^{-1}$) are required when dealing with precipitates formed by organic compounds of high molecular weights (16a).   The chief problem lies in adequately linking the

precipitation system to the atomic absorption (AA) spectrophotometer instrument because direct cation measurement falls outside
the linear range of instrument. Our team has overcome this problem (15-17) by continuous dilution of the precipitating cation with
a water stream prior to introduction into the spectrophotometer.
This entails using the carrier containing the precipitating cation
at a lower flow rate (between 0.7 and 3 mL min^{-1}) than the water
stream used for dilution, which is circulated at a rate of 2.5 to
5.0 mL min^{-1}. The ideal flow rate ratio between both streams is
a function of the AAS sensitivity of the particular metal. Obviously, cation dilution is unnecessary when the precipitate is dissolved prior to measurement.

2.3.4 Influence of FIA Variables on Precipitation Efficiency

The efficiency of a continuous precipitation system is influenced directly by the injected sample volume and the geometric characteristics of the precipitation coil. The higher the injected sample (nFIA)
or reagent (rFIA) volume is, the larger the amount of precipitate
formed and the greater the analytical signal will be. Yet the increase in analytical signal with the injected volume is concomitant
with an increase in the blank signal, so that the peak height increments are virtually constant over a wide volume range (50 to 400 µL).
Moreover, the peak width does increase—and sample throughput decreases—with increasing amounts of precipitate formed. This, together with the fact that high injected volumes yield irreproducible
results, have led authors to use volumes between 50 and 140 µL in
the experiments carried out to date (15-17).

The FIA variable affecting precipitation efficiency to the greatest extent is the coil length. All the studies performed on this subject have led to the same conclusion; namely, the occurrence of
three distinct zones (Fig. 4)

1. The *first zone* corresponds to short lengths, usually less than
   100 cm. The plug's residence time is so short that mixing is
   inadequate and the reaction does not proceed to completion, so
   that the plug still contains unprecipitated cation by the time
   that measurements are made.

2. The *second zone* comprises a length range resulting in virtually
   complete precipitation and practical absence of free cation. It
   is obviously the zone of greatest efficiency.

3. The *third zone* corresponds to a residence time longer than that
   needed for full precipitation, after which in nFIA the precipita-
   tating cation diffuses mainly through the tall interface into the

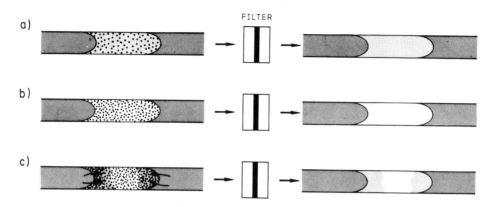

FIGURE 4 Influence of the precipitation coil length [(a) short; (b) medium; (c) long] on the reaction zone in the flow, prior to and after the filter.

precipitate plug, which thus yields a signal corresponding to the diffused cation. Inasmuch as the extent of this phenomenon increases with increasing coil length, it is inadvisable to use lengths longer than 300 cm, which result in substantially decreased signal increments in the precipitation process (15).

Only the first two zones occur in rFIA as a result of the carrier containing no cation that might diffuse into the precipitate plug. Other coil variables, such as the inner or turn diameter, had little or no influence according to these authors.

2.3.5 Sensitivity

Continuous precipitation systems considerably increase the sensitivity of methods developed by conventional techniques. For a given chemical system and precipitating cation concentration, the sensitivity is greater when the cation is inserted into a sample carrier stream (rFIA) than when the sample itself is injected (nFIA). This is probably due to the fact that when the sample is used as carrier and the reagent is injected into the flow, the amount of sample in the reagent zone increases with increasing dispersion, thereby resulting in a greater extent of mixing and precipitation at lower sample concentrations (16, 16a). No significant differences with respect to the nondissolution technique are observed, though, when the precipitate formed upon sample injection (nFIA) is dissolved prior to measurement (15).

2.3.6 Selectivity

As shown in many of the papers quoted in the latest monograph on
FIA (3), this technique usually results in more or less substantially
decreased interferences compared to conventional methods. This is
also the case with FIA precipitation systems. Thus our team (16)
has shown that the determination of chloride ion by precipitation
with silver on a nondissolution precipitation manifold is more tolerant of potential interferents than is its manual counterpart. This
increased selectivity can be further enhanced by dissolving the precipitate prior to determinating the analyte, since the species copre-
cipitating with silver may not be soluble in the reagent used to dissolve the precipitate. Such is the case with iodide ion; this forms
an insoluble halide with silver which, unlike that of chloride, is
ammonia insoluble, which allows the determination of chloride in the
presence of high iodide concentrations. Table 1 lists the selectivity
levels achieved in the determination of chloride by precipitation with
silver nitrate with and without precipitate dissolution in an FIA system and compares them with those afforded by the nondissolution
manual technique. (The selectivity factor is defined as the ratio
between the tolerance ratios for a given foreign species in the determination of the same analyte by two different methods.)

The greater tolerance of FIA precipitation methods—and, in
general, of FIA methods—to foreign species can be attributed to
their kinetic nature (4). Thus the results obtained in the interference study allowed the authors' team to conclude that the increased selectivity of continuous precipitation methods is a result
of the absence or scarce occurrence of undesirable side reactions
during the short measuring interval (residence time in FIA). In
manual precipitation methods, which entail waiting for the complete
decantation of the precipitate before it is centrifuged or filtered,
the interferents and the precipitating cation are in contact for a
longer time—at least tenfold—than in FIA, so that there is an increased risk of loss of the former through reaction with one or
several of the foreign species present.

2.3.7 Filter Cleaning

As the most important device in a precipitation system, the filter
should be cleaned periodically to remove the precipitate formed after
a number of analytical determinations. Our team (15-17) has carried
out a comprehensive study on the potentially adverse effect that
precipitate buildup on the filter may have on the analytical signal.
By way of example, a filter with a chamber inner volume of 580 µL
and a filtration area of 3 cm^2 should be cleaned every 250 samples
for curdy or gelatinous precipitates and every 20 samples for crystalline precipitates when the nFIA mode is applied if reproducible

TABLE 1   Comparison of the Selectivity Level
Achieved in Indirect Atomic Absorption Deter-
mination of Chloride by Precipitation with Ag(I)
by the Conventional Procedure (1) and in Flow
Injection Manifolds Without (2) and With (3)
Precipitate Dissolution

Foreign ion	Selectivity factor	
	2/1	3/1
I^-	2.0	340.0
Br^-	7.0	70.0
$Fe(CN)_6^{4-}$	6.0	20.0
IO_3^-	6.0	15.0
ClO^-	12.0	14.0
$Fe(CN)_6^{3-}$	10.0	10.0
CN^-, SCN^-	8.0	8.0
SO_4^{2-}	6.0	6.0
AsO_4^{3-}, BrO_3^-, AsO_2^-	5.0	5.0
CrO_4^{2-}	4.0	4.0

Source: Reproduced with permission of the
Royal Society of Chemistry (4).

results are to be obtained (Fig. 5).   The filter cleaning frequency
roughly doubles when it is the precipitating cation that is injected
into the sample stream (rFIA).   Filters and columns should ideally
be cleaned for 5 min with the same reagent as that used for precipitate dissolution in ultrasonic baths.

The difference between the number of consecutive determinations afforded by curdy or gelatinous precipitates and crystalline
precipitates makes it advisable to use colloid rather than crystalline
precipitates in FIA, unlike conventional manual techniques.   This is
a result of the greater compactness and lesser permeability of crystalline precipitates, which occupy the filter pores, thereby clogging
the filter outlets and eventually giving rise to system breakdown.
Gelatinous and curdy precipitates, which are less compact, allow
for readier passage of the flowing solutions and therefore call for
less frequent filter cleanups.   On the other hand, continuous precipitation systems with precipitate dissolution require no filter
cleaning whatsoever.

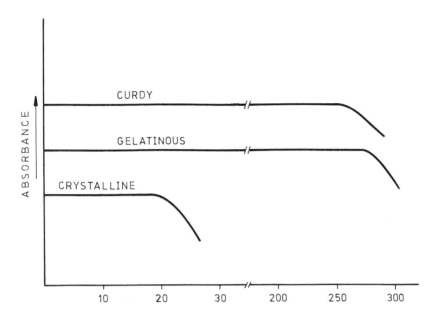

FIGURE 5 Variation of absorbance as a function of the number of identical samples injected for different types of precipitates. [Reproduced with permission of the American Chemical Society (15).]

2.4 Determination of Anions

Precipitation systems and FIA have rarely been used in conjunction so far, probably because of the still short life of this flow technique. One interesting application among the few reported so far is the indirect determination of oxalate, chloride, and iodide developed recently.

The determination of oxalate is based on the manual procedure developed by Menaché (18) for the analysis for oxalic acid in urine. The method involves precipitating the oxalate in urine (between 17 and 186 μg mL^{-1}) with excess calcium at pH 5 and determining non-precipitated calcium in the filtrate by AAS. This is quite a slow procedure, as it entails allowing the samples to cool down for 24 h to ensure complete precipitation of calcium oxalate, which results in a sample throughput of only 30 samples per day. By use of a continuous precipitation system, the authors' team (15) carry out this determination by injecting 50 μL of solutions containing between 5 and 90 μg mL^{-1} oxalate into a 60 μg mL^{-1} solution of Ca^{2+} in acetic-acetate buffer. Owing to the sluggish precipitation, the sample plug must be stopped at the reactor for 1 min and be heated at 60°C. Unlike the manual method, this is quite fast, as it allows

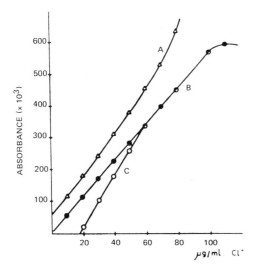

FIGURE 6   Calibration curves for chloride obtained with different
concentrations of nitric acid as a washing solution: (A), (B), and
(C) for 1×10^{-3}, 5×10^{-3}, and 10×10^{-3} M of nitric acid, respectively; dissolving solution, 6 M NH_3.   [Reproduced with permission
of the American Chemical Society (15).]

one to carry performance up to 20 determinations per hour.   The
precipitate, once retained in the filter, is dissolved with a 0.2 M
solution of EDTA in 7 M ammonia after washing with distilled water.
The determination range is somewhat narrower than that of the nondissolution procedure, owing to the difficulty involved in dissolving
the precipitate for injected oxalate concentrations above 60 µg mL^{-1}
(this shows in the appearance of a plateau rather than the traditional FIA peak).   The procedure involving precipitate dissolution,
which—in principle—should be more accurate than the nondissolution
method because of the intermediate washing operation included to remove any cation vestiges adsorbed on the precipitate, has in fact a
higher relative standard deviation (9% versus 5%).   This can be attributed to the greater number of steps involved in the procedure
with precipitate dissolution, which also results in decreased sampling frequency.

    With a similar assembly, our team has developed a method for
determination of chloride (16).   The sample containing the halide is
inserted into a stream of 80 µg of Ag$^+$ per milliliter.   Precipitation

TABLE 2 Analytical Features of the Procedures for Determination of Chloride by Various Methods

Parameter	nFIA[a]	nFIA[b]	rFIA[a]	Batch[a]
Linear range (μg mL^{-1})	3-100	3-100	0.3-10	5-100
Detection limit (μg mL^{-1})	1.3	2.1	0.1	2.0
Relative standard deviation (%)	2.0	5.2	3.5	4.7
Sampling frequency (h^{-1})	50	10	200	5

[a]Without precipitate dissolution.

[b]With precipitate dissolution.

Source: Reproduced with permission of the Royal Society of Chemistry (16).

is instantaneous and thus calls for relatively short reactor lengths. The sensitivity of the method is similar to that of its manual counterpart. Once retained on the filter, the precipitate can be washed with dilute nitric acid and then dissolved in 6 M ammonia (15). To ensure a thorough cleanup of the precipitate, the authors ran a series of calibration curves at different nitric acid concentrations between 1×10^{-3} and 10×10^{-3} M (Fig. 6). As can be seen, the use of a 1×10^{-3} M solution [Fig. 6(a)] results in incomplete washing, so that the calibration curve obtained after the precipitate dissolution has a nonzero intercept. A nitric acid concentration of 10×10^{-3} M [Fig. 6(c)] ensures complete washing, but also dissolves precipitates obtained from chloride concentrations below 20 μg mL^{-1}, which therefore yield no signal after subsequent dissolution with ammonia. A concentration of 5×10^{-3} M [Fig. 6(b)] thus seems to be the best compromise to ensure thorough washing without the risk of premature precipitate dissolution. The sensitivity of the procedure with and without dissolution is very similar in this case (see Table 2).

The high baseline recorded upon injection of chloride into a silver stream (approximately 0.6 absorbance unit for silver signal) and its associated oscillations have led the authors to develop an alternative rFIA determination (16). The chemical variables involved in this (optimum silver concentration, pH, etc.) do not differ much from those of the nFIA mode; yet the FIA variables are markedly divergent. Thus the rFIA procedure affords higher carrier (sample) flow rates, larger injected volumes, and shorter reactor lengths. As far as silver is concerned, its injection into the

carrier gives rise to a smooth baseline—it is water or the chloride
sample which passes through the detector. In addition, its sensitivity and sampling frequency are tenfold and fourfold, respectively, those of the nFIA procedure, and less frequent filter cleanups
are required (roughly one every 400 samples). Table 2 summarizes
the analytical features of the determination of chloride by FIA methods and a batch method which are applied to the determination of
the halide in various types of water (drinking, well, spring, and
river water).

By use of an nFIA configuration identical with that employed
in the determination of chloride, the authors have developed a procedure for determination of iodide over a molar concentration range
similar to that achieved for chloride (17).

2.5 Determination of Mixtures

There are by now a host of simultaneous FIA determinations of different nature reported in the literature (19). Simultaneous determinations based on the use of a continuous precipitation-dissolution
system can be implemented merely by finding two species precipitating jointly with the same cation, provided that one of the precipitates obtained is selectively soluble in a given solvent. This novel
application of FIA precipitation systems is of great interest, as so
far, there are no literature references to simultaneous indirect AAS
determinations involving precipitation reactions.

Our team has recently proposed the sequential determination
of chloride and iodide based on the different solubility of their silver precipitates in ammonia (17). The sample containing both halides is injected once into a main line carrying an 80-μg mL^{-1} Ag$^+$
solution. Once both silver halides have precipitated and the precipitate has been retained, the overall halogen concentration in the
sample is determined using water as a blank. In a second step,
the precipitate is washed with 5×10^{-3} M nitric acid to remove adsorbed silver, and then silver chloride is dissolved selectively in
6 M ammonia (preliminary assays showed the silver iodide precipitate to be insoluble in the ammonia used for iodide concentrations
in the range 10 to 320 μg mL^{-1}). The signal obtained from the silver in the chloride is directly proportional to the concentration of
this halide in the sample. The difference between the concentrations found in both steps corresponds to the iodide concentration
in the sample. The method has been applied to the resolution of
mixtures of these two anions in a wide concentration ratio range;
it has also been applied to a variety of foods, including wines,
milk, cheese, bread, coffee, and eggs. Inasmuch as these contain
no iodide or concentrations below the detection limit of the method,
its application requires adding this halide to the samples to be

assayed. The results obtained are consistent with those provided
by the standard potentiometric method. The sampling frequency
afforded is 10 h^{-1} and the relative standard deviation is less than 6%.

2.6 Determination of Pharmaceuticals

The vast potential of the association of FIA and continuous precipi-
tation-filtration systems has recently been shown in the determination of pharmaceuticals. As a rule, organic pharmaceuticals are determined by reaction with metals forming either a precipitate or an
extractable ion-association compound or other complex, the metal
content of which is measured by AAS (5). Our team is currently
working on the application of a continuous precipitation system to
the analysis for active principles in pharmaceutical preparations;
one of this work's fruits has been the development of an indirect
determination for sulfamides based on their precipitation with silver
nitrate (16a) and the use of an rFIA configuration [Fig. 2(b)] where-
in the silver salt (200 µg of Ag$^+$ per milliliter) acts as precipitating
the silver salt (200 µg of Ag$^+$ per milliliter) acts as precipitating
agent. The pH of both the carrier and the precipitating solution
is critical and should be in the range 8 to 9 and 6 to 7, respectively. A thorough study of the FIA variables involved revealed that
(a) precipitation coil lengths between 50 and 400 cm can be used,
(b) the optimum injected volume is about 100 µL, and (c) the water
dilution/carrier flow rate ratio should be greater than 3, owing to
the high silver concentration required. Unlike its manual counterpart, which requires 100 mg of sulfamide and a large excess of silver, the FIA method is quite sensitive (linear concentration range
between 3 and 35 µg mL^{-1}) and has a detection limit of 1 µg of sulfamide per milliliter. In order to make the determination applicable
to as wide a range of pharmaceutical preparations as possible, the
authors studied the influence of various excipients and diluents
commonly found in them: namely, vanillin, polyvinylpyrrolidone,
glucose, lactose, sucrose, glycerol, starch, ethylene glycol, and
so on. No matrix effect arising from the presence of any of these
substances was found. The method is applicable to the determination of sulfamides in tablets, injectables, drops, and so on.

2.7 Preconcentration of Traces

A novel use of continuous precipitation systems—trace preconcentration—has been reported by the authors (20). So far, trace preconcentration was generally carried out by coprecipitation with a carrier precipitate (collector), as direct separations based on precipitation reactions had as a disadvantage the small amounts of precipitate
formed and hence their cumbersome manipulation. However, the
collector added may still be present in the dissolved precipitate at

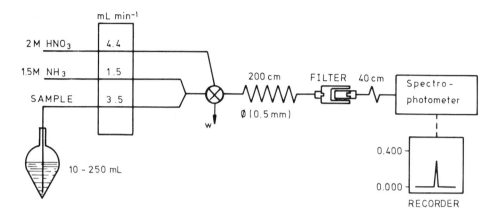

FIGURE 7 Schematic diagram of the configuration designed for the preconcentration of lead traces in waters. [Reproduced with permission of the Royal Society of Chemistry (20).]

concentrations well above those of the cations to be determined and should therefore be separated prior to determination of the traces of interest. This preconcentration technique requires large sample volumes, is rather time-consuming (i.e., results in low sample throughput), and occasionally provides irreproducible results arising from the manipulation preceding the analytical measurements (11,12). On the other hand, the technique proposed calls for no collector, as direct separation is feasible however small the amount of precipitate obtained may be, because of the absence of manipulation in continuous precipitation systems.

After a comprehensive study of the variables involved, the authors proposed the setup depicted in Fig. 7 for the preconcentration and subsequent determination of lead in samples containing between 0.3 and 15 µg of Pb^{2+} in volumes between 10 and 250 mL continuously aspirated into the system and mixed with the precipitating reagent (1.5 M ammonia). Precipitation is instantaneous and the basic salt formed is retained on a filter of suitable size. The precipitate is subsequently dissolved with an acid stream (2 M HNO_3) and the transient signal yielded by dissolved lead is measured by a spectrophotometer. The precipitate requires no washing prior to its dissolution. The method affords a concentration factor of up to 700. The detection limits achieved a range between 1 and 20 ng mL^{-1}, depending on the sample volume used, and the relative standard deviation is always less than 3.6%. The selectivity is also quite good; determination is interfered with only by those anions precipitating lead as salts insoluble in 2 M nitric acid (e.g., sulfide) or by cations precipitating with ammonia [e.g., Sn(IV) and Fe(III)] and not readily soluble in the nitric stream into which the lead basic salt may be occluded. Interference from sulfide and iron can be overcome by adding 3 mL of 30% hydrogen peroxide and 2 mg of tartrate, respectively. Cations forming ammonia complexes are tolerated at concentrations 400 times as high as that of lead, while anions such as nitrite, nitrate, carbonate, sulfate, chloride, bromide, iodide, and phosphate pose no problem even at concentrations 500 times higher. The method has been applied satisfactorily to the preconcentration and determination of lead in drinking water over the concentration range 1.2 to 60 ng mL^{-1}. Although it has so far been applied only to lead, this technique for preconcentration of metal traces can indeed be extended to a host of other metals.

2.8 Filterless Continuous Precipitation Systems

Růžička et al. (14) have recently developed a novel flow injection system for conversion of soluble species into insoluble compounds in which the precipitate is directly led to the spectrophotometer nebulizer. Originally proposed for the AAS determination of sulfide, it is based on the precipitation of this anion with Cd(II) as precipitating tag reagent. Sample and reagent are injected simultaneously following the merging-zone principle and the excess reagent (cadmium) is retained on an ion-exchange column packed with a chelating resin (8-quinolinol immobilized on controlled-pore glass).

The system has been automated by means of an injection valve, a dual-pump setup, and an electric timer synchronizing the stop and start of both pumps (Fig. 8). Each cycle is started by injecting 200 µL of sample and 250 µL of the tag reagent into two separate carrier streams (25 mM ammonium acetate buffer of pH 9.5) propelled forward by pump P$_2$. Free cadmium is completely retained on the column; cadmium sulfide passes through the column into the instrument. Once cadmium has been sensed and the recorded signal has reverted almost to baseline level, the valve (V) is switched manually to the elution position, which prompts the trigger to start pump P$_1$ and stop P$_2$. During this second stage, all cadmium(II) retained on the ion-exchange column is redissolved by the eluent (1 M HNO$_3$) and transferred to the nebulizer, where it yields a second peak. Simultaneously with the elution procedure, fresh sample and reagent solutions are aspirated into the valve loops.

The choice of cadmium(II) as precipitating agent is justified by the low solubility of its sulfide, its good AAS sensitivity, and the lack of anions forming insoluble compounds with it. It should be noted that the injected sample volume used is greater than that of reagent in order that the sample zone be fully wrapped in that of cadmium and no free sulfide may reach the column; otherwise,

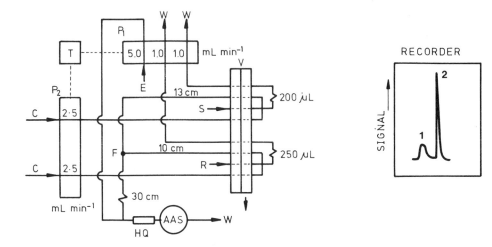

FIGURE 8   Flow-injection manifold for the determination of sulfide
with detection by AAS, comprising two peristaltic pumps (P_1 and
P_2); a timer, T; an injection valve, V; and an ion-exchange column,
HQ. S, sample; C, carrier stream; R, reagent; E, eluent; W, waste.
Recordings: 1 and 2, signals corresponding to the cadmium sulfide
and elution of excess of cadmium(II). [Reproduced with permission
of Elsevier Science Publishing (14).]

sulfide might react with cadmium(II) withheld on the column and be
retained within the structure of the complexing material, thus lowering the resin exchange capacity, increasing flow resistance, and
ultimately giving rise to poorly reproducible results. According to
the authors, the best results as far as the resin is concerned are
provided by 8-quinolinol immobilized on controlled-pore glass.
     Of the 14 potential interferents tested by the authors, only
phosphate, forming a compound of rather low solubility with cadmium (i.e., tricadmium diphosphate), was found to be a serious disturbance. This can, however, be overcome by adding excess ortho-
phosphate (200 μg mL^{-1}) to the samples prior to their analysis.
This results in slightly increased sensitivity and hence in a somewhat narrower determination range: 0.4 to 2.4 μg mL^{-1}.
     Despite the good sensitivity achieved, the authors do not pro-
vide a systematic study of the chemical (pH, optimum precipitating
reagent concentration) and FIA variables (injected volumes, flow
rates, coil length) involved. This and another detailed study of
potential cationic interferences liable to precipitate with sulfide and/
or be retained on the column (the work mentions only Ca^{2+} and
Mg^{2+}) might have assisted in evaluating the selectivity of the method.

This conversion technique for determination of anions could be simplified considerably by using an on-line filter, which would dispense with the separation of unreacted reagent by means of the ion-exchange column. On the other hand, it is widely accepted that samples to be inserted into the AAS flame should be liquid; consequently, solid samples should be appropriately dissolved prior to introduction into the nebulizer (21). Despite the foreseeable errors and decreased nebulization efficiency that one would expect from the direct insertion of precipitates into the flame, the method boasts amazing precision (relative standard deviation less than 1.3%) probably resulting from the colloid nature of the cadmium sulfide precipitate and the small amount reaching the nebulizer, in addition to the ease of oxidation of the cation. Precipitates of a different nature would probably pose major problems arising from thermal instability.

3 LIQUID EXTRACTION

3.1 Basic Components of a Continuous Liquid-Liquid Extraction System

Every continuous liquid-liquid extraction system invariably has three essential components: the solvent segmentor, the extraction coil, and the phase separator. The *solvent segmentor*, where the organic and the aqueous phase converge, functions to provide alternate regular segments of both phases prior to entering the extraction coil.

The segmentor most frequently used in continuous liquid-liquid extractors coupled to AAS spectrophotometers is a modification of the Technicon A-8 and A-10 connectors (3). Some authors use special homemade segmentors. Bengtsson and Johansson (22) have devised one very similar to the Technicon A-10 but with a confluence point featuring a smaller dead volume. Swelleh and Cantwell (23) use a segmentor made by drilling out the small-bore cylindrical chamber of a commercially available Kel-F tee and inserting 2-mm-long flared pieces of 0.8-mm-ID Teflon tubing into the three branches of the tee.

The transfer of matter between the segments of both immiscible liquids takes place in the *extraction coil*. Because of the corrosive properties of organic solvents, this element should preferably be made of Teflon or glass. According to Shelly et al. (24,25), if the analyte is to be originally contained in the aqueous phase, the coil used should be of Teflon in order that the aqueous phase, repelled by the tube walls, be carried as bubbles, thus ensuring as efficient an extraction as possible. Since all chemical systems commonly used in FIA-AAS extraction processes involve placing the

analyte in the aqueous phase, it is not surprising that Teflon coils
have been used almost exclusively in this area. The tubing inner
diameter usually ranges between 0.5 and 1.0 mm, and their length
between 2 and 3.6 m. Most of them afford an extraction efficiency
in the range 70 to 90%, depending on the system variables—particularly on the flow rate ratio between both phases and the geometric
characteristics of the coil.

The *phase separator* is intended to split the incoming segmented
flow continuously into two streams, each containing one of the phases.
Although the separation efficiency usually achieved ranges from 80
to 95%, quantitative separation of the phases is required only when
back-extraction is involved, since the analyte concentration in the
organic phase is independent of the recovered volume. However,
the phase in which the determination is to be carried out should be
kept completely free from the other phase to avoid anomalous signals
on passage through the detector.

There are two basic types of phase separator, T-shaped and
membrane separators. The T-shaped separator most commonly used
in FIA-AAS is the Technicon A-4 model (26-29). Membrane separators are far more common. The membrane is usually made of micro-
pore Teflon (pore size 0.5 to 1.0 μm), which allows selective passage
of aqueous phase-free organic phase. All the continuous membrane
separation units reported so far have been designed by the authors.
They are very similar and are based on the basic construction proposed by Nord and Karlberg in 1980 (30). They consist of two
drilled Teflon or Perspex blocks retaining the membrane in the separation minichamber. Figure 9 depicts one such separator. The
incoming (segmented) and outgoing aqueous phase streams, which
contain some organic phase, run in the same direction, while the
outgoing organic phase stream (MIBK), lighter than the aqueous
phase, is always on top of this and emerges through a hole drilled
at the center (31,32) or at one of the ends (33,34) of the separation minichamber. Other phase separators include a chamber with
an angled PTFE surface as well as a membrane (35).

According to Nord and Karlberg (30), the volume of the separator groove should be at least four to five times as large as the
volume of an individual segment of aqueous phase since the former
should not be filled by the latter alone. Taking into account that
the typical volume used in FIA extractions is about 3 μL, the minimum volume of a separator should thus be 15 μL or larger. The
separation minichamber volume usually ranges between 36 and 60 μL.

Our team has carried out a comparative study (31) of both
types of separator and has shown the membrane separator depicted
in Fig. 9 to result in better extraction efficiency than the A-4T
phase separator—only 79% as efficient as the former. A miniature
extraction module, including engraved conduits for mixing of sample

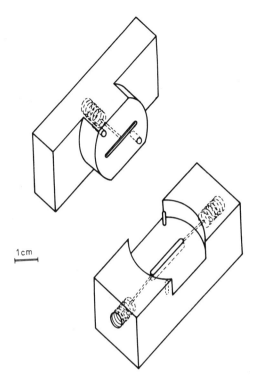

FIGURE 9 Drawing of a membrane phase separator. [Reproduced
with permission of the American Chemical Society (31).]

and reagent, an engraved segmentor, a detachable extraction coil,
a membrane separator, and a rinsing system for the flow cell has
been reported (36).

Although not essential components of continuous liquid-liquid
extraction systems, *liquid displacement bottles* are rather common-
place. These reservoirs are usually filled with an organic solvent
with or without a reagent. By pumping a water stream into the
bottle, the solvent is forced out of the bottle onto the segmentor
at a regular flow rate.

Direct pumping of methyl isobutyl ketone (MIBK) with a peristaltic pump poses a still unsolved problem arising from (a) the
difficulty involved in achieving regular flow rates owing to deterioration of the pump tubing (e.g., Viton rubber in Acidflex) caused
by the attack of the organic solvent, and (b) clogging of the membrane pores by small particles rubbed off the inside of the pump
tube (32). These shortcomings can be circumvented by keeping
the organic solvent out of contact with the pump tubes or by using

constant-pressure pumping. Sweileh and Cantwell (23) use pressurized gas nitrogen to propel both the MIBK stream and the remainder of the aqueous streams involved in the extraction assembly. Bäckström et al. (26) use Solvaflex tubing to propel the solvent employed in their extraction system (Freon 113).

3.2 Theoretical and Practical Considerations

3.2.1 Organic Solvents

Methyl isobutyl ketone is by far the extractant most commonly used in this field; in fact, there are only two precedents (22,26) for the use of Freon 113, which requires back-extraction prior to introduction into the instrument. The obvious advantages and few disadvantages of the use of MIBK in AAS have been emphasized by Cresser (37). Other solvents used in conventional extraction procedures are banned from automatic extraction since their shortcomings clearly surpass their advantages. The FIA technique, which uses low solvent volumes (about 100 µL) injected into water streams carrying them direct to the nebulizer (33), allows for use of a variety of solvents otherwise unaffordable because of the toxic products formed by combustion (37). Tyson et al. (38) have shown that the injection of 70 µL of a lead complex extracted in carbon tetrachloride into a water carrier prior to nebulization results in a phosgene concentration in the instrument's environment of less than 0.05 µg mL^{-1} (the TLV of phosgene is 0.1 µg mL^{-1}).

Organic solvents are also used as carriers to increase the sensitivity of the determination of various elements in systems where no proper liquid-liquid extraction takes place. Fukamachi and Ishibashi (39) inject aqueous solutions of various metals into methanol, MIBK, and *n*-butyl acetate, the last two of which give rise to more enhanced signals. Attiyat and Christian (40) use water, methanol, ethanol, acetone, and MBIK as carriers in the determination of copper. Unlike the Japanese authors, they inject cation solutions into one of these solvents or a mixture of them and have studied in greater depth the dispersion in the mixing coil and the FIA variables involved. Neither the former nor the latter authors use pumps as propelling devices; in fact, this role is allocated to the nebulization system itself.

3.2.2 Linkage of the Nebulizer Feed System to
the Extraction System

Most commercial atomic absorption instruments are designed for sample feed rates in the range 3 to 10 mL min^{-1}, which result in sensitivities below 3 mL min^{-1}. Obviously, a concentration ratio of aqueous to organic phase of 10 will require a minimum sample flow rate of 30 mL min^{-1} to keep the organic flow rate above 3 mL min^{-1}. Hence direct continuous feeding of the nebulizer with organic extract is technically impractical. Two configurations for integration of a low-flow-rate extraction system and a high-flow-rate AAS nebulizer feed system have been proposed so far. In the configuration *without sample injection* devised by Nord and Karlberg (33), the sample is continuously introduced into the extraction system. The organic phase stream emerging from the separator at a flow rate below 1 mL min^{-1} is linked to an injector loop, the content of which is unloaded into a water stream, leading it to the AAS instrument at a flow rate of 4 to 5 mL min^{-1} so as to ensure the best possible performance from the spectrophotometer. Integration is thus accomplished with an insignificant loss of sensitivity. In the configuration *with sample injection*, it is always the sample that is injected in the system. The organic phase stream emerging from the separator at a flow rate between 1 and 2 mL min^{-1} is accelerated prior to reaching the nebulizer either with the aid of a T-piece (35) placed between the separator and the nebulizer to match the flow rate to that required by the nebulizer in the air compensation method (41), or alternatively, by merging the organic phase with an MIBK stream from an open flask at a three-way connector (23). Thus the difference between the nebulizer aspiration rate (6.0 mL min^{-1}) and the flow rate of MIBK across the membrane (2.1 mL min^{-1}) is compensated for by the additional stream of pure MIBK. The use of air as compensating fluid results in rougher baselines and irreproducibility in the AAS signal. The two techniques used to compensate the divergence between the flow rate of the organic phase extract stream and the nebulizer aspiration rate are illustrated in Fig. 10.

3.2.3 Influence of FIA Variables on Extraction Efficiency

The efficiency of an extractive procedure is especially influenced by the geometric characteristics (length, inner diameter, turn diameter) of the extraction coil, the flow rate of the aqueous and organic phase and the ratio between the two phases, and the length of the fluid segments.

The optimum extraction coil length to be used when the sample is introduced in a continuous manner is between 200 and 500 cm (27-29,31,34). For a given volume a roughly constant efficiency is obtained for coil inner diameters between 0.5 and 0.7 mm and turn diameters up to 60 mm (27-29,31). The FIA variables most markedly affecting extraction efficiency are no doubt the flow rates of both phases and their ratio. The higher the ratio of the aqueous phase flow rate to that of the organic phase, the more efficient the extraction will be as a result of the analyte concentrating in the organic phase. Yet this increasing trend is limited by two factors: too high flow rates result in irregular segmentation between both phases (27,34), and MIBK is scarcely soluble in water (31,34).

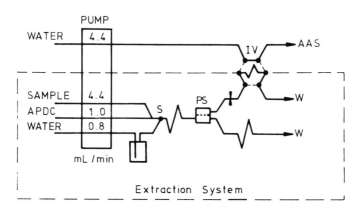

(a)

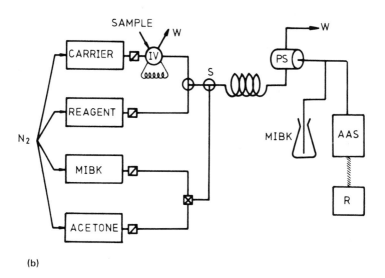

(b)

FIGURE 10 Manifolds for extraction of aqueous samples (a) without
injection and (b) with injection of the sample. IV, injection valve;
S, segmentor; PS, phase separator; W, waste; R, recorder. [Re-
produced with permission of Elsevier Science Publishing (33) and
the American Chemical Society (23).]

Despite these two limitations, a large variety of systems involving very different flow rate ratios have been reported. Indeed, membrane separators afford higher flow rates than do T-separators (31). The overall flow rate employed in all the extraction assemblies reported to data never exceeds 5 mL min^{-1}. The length of the aqueous-organic fluid segments does not affect significantly the efficiency of extraction systems involving continuous introduction of the sample, although it does have some effect on it when the segments produced are somewhat shorter (42).

There are only two unsegmented liquid-liquid extraction manifolds involving sample injection (23,35). The influence of only one of the variables—the coil length—on their extraction efficiency has been studied. In any case, such efficiency will increase with decreasing sample dispersion. The overall flow rate set in both manifolds is between 3 and 10 mL min^{-1}.

3.2.4 Dispersion in Flow Injection Extraction Systems Coupled to AAS Instruments

Although most of the factors governing dispersion in single-phase flow injection systems are well known and have been thoroughly described by Růžička and Hansen (2), dispersion in two-phase flow injection systems is still a relatively unexplored subject. Nord and Karlberg (42) have investigated the mechanism of dispersion in two-phase systems. These authors assume that the phase with greater affinity to the tubing material forms a thin film on the walls and the analyte to be retained in this stationary film and is then partly mixed into the next segment by diffusion and convection, so that some analyte lags behind the sample zone, thereby resulting in dispersion, the extent of which increases with increasing linear flow velocity and tubing inner diameter. They have also observed that dispersion is minimal in systems where the phase carrying the analyte does not form the aforementioned film. Yet the mechanism representative of dispersion in flow injection extraction systems proposed by these authors is based on data obtained from the pentanol-water solvent, where the organic solvent does form a thick film on the tubing walls, thereby favoring dispersion. Obviously, the mechanism cannot be extended to extraction systems involving organic solvents forming thin or no films. No analyte dispersion in the extraction coil occurs in continuous extraction systems, where the sample is introduced in a continuous manner (i.e., most applications of the AAS technique) (42).

3.2.5 Sensitivity

Solvent extraction is a means of substantially increasing the sensitivity of AAS methods through preconcentration and improved flame atomization (37). Flow injection liquid-liquid extraction methods are somewhat less sensitive than manual techniques using comparable flow rate ratios. This lower sensitivity is the result of (a) a shorter reaction time giving rise to partial extraction, and (b) physical dispersion or dilution of the sample in the carrier. These are opposing effects as any attempt at increasing the reaction time (hence favoring the reaction/extraction) by use of longer coils or lower flow rates invariably results in increased dispersion, and vice versa.

Of the two FIA extraction modes, that involving continuous insertion of the sample and injection after extraction provides greater enhancement factors than that involving sample injection—even for injected volumes of 1 mL—because of increased sample dispersion in the latter. By injecting Cu(II) into an APDC solution and using MIBK as solvent, Tyson et al. (38) obtained a calibration curve with a slope 1.3 times as steep as that of the corresponding aqueous phase calibration and half that of the conventionally nebulized aqueous phase for the injection of 1000 and 50 μL, respectively. The sensitivity is increased by a factor of 15 to 20 for a system based on the same chemistry when the Cu(II) samples are introduced in a continuous manner (34).

3.2.6 Selectivity

The increased selectivity resulting from the use of continuous liquid-liquid extraction prior to the determination can be exploited in two ways: as a means of avoiding spectral interferences from other species present in the sample by selective extraction of what is of interest (thermodynamic discrimination), usually by a complex-formation reaction, or by kinetic discrimination (i.e., on the basis of the difference between the rate of reaction of the analyte and those of the potential interferents). Sweileh and Cantwell (23) have shown that it is possible to overcome spectral interferences by use of a liquid-liquid extraction system. By way of example, these authors determine zinc at a wavelength of 213.9 nm, coincident with the absorption wavelength of iron, which is therefore a serious interferent. By use of an extraction manifold in which Zn(II) is extracted as thiocyanate into MIBK, these authors determine between 0.5 and 1.6 μg of zinc per milliliter in synthetic samples containing 2% iron, with errors below 6%. The same determination yields errors between 40 and 80% when implemented by AAS. However, no comparison is made by these authors between the FIA method and its manual counterpart, which would probably not be so divergent.

Gallego and co-workers (31) have reported a comparative study of the interferences found in the indirect determination of surfactants by formation of an ion pair between the surfactant and 1,10-phenanthroline/copper(II), with both continuous and manual extraction. As can be seen from Table 3, the FIA method is more selective

TABLE 3 Comparison of the Tolerance to Interferents of the FIA Technique and the Batch Procedure Using the Copper-Phenanthroline Method

Foreign ion	Tolerated ion-to-LAS[a] ratio	
	FIA method	Batch method
$PO_4{}^{3-}$, $SO_4{}^{2-}$	1000	1000
$NO_3{}^-$	1000	<5
Br^-	1000	50
Co^{2+}	1000	<50
Phthalic acid	1000	50
Succinic acid	1000	50
Glutamic acid	1000	50
Benzoic acid	500	50
Ni^{2+}	250	<50
Triton X-100	250	5
I^-	25	10
$ClO_4{}^-$	25	<5

[a]Sodium dodecyl sulfate.

Source: Reproduced with permission of the American Chemical Society (31).

than its manual counterpart. This greater selectivity is accounted for by the authors on the basis of the fact that the liquid-liquid partition equilibrium is not reached in the FIA system, so that as the experimental conditions are suited to the analyte determination, the potential perturbation of the other species is minimized by the short contact time between phases. Thus the increased selectivity of the FIA procedure should apparently be attributed to a kinetic factor.

3.3 Determination of Metals

Liquid-liquid extraction is the separation technique most frequently used in conjunction with AAS to lower the detection limit of metal ions. Yet few methods using this technique in combination with an FIA-AAS system have been proposed so far. The first manifold exploiting such an association was used by Nord and Karlberg in 1981

(33) for the determination of Cu(II) based on the automatic extraction of the metal into MIBK, using ammonium pyrrolidinedithiocarbamate (APDC) as extractant. The sample was continuously pumped and mixed with a stream of APDC, segmented by the MIBK stream with the aid of a segmentor, and extracted in the extraction coil. A fraction of the organic phase was separated by the membrane phase separator and led into the injector, located in an integrated feed system of the atomic absorption spectrometer. Karlberg's determination has two shortcomings. One is the narrow linear range resulting from precipitation of the APDC-copper complex in an aqueous medium at concentrations above 0.15 µg of Cu(II) per milliliter before the sample is introduced into the extraction system. The resultant absorbance interval is so narrow (0.08 to 0.080) that FIA variables can be studied only by using Cu-MIBK standards. The second problem lies in the fact that contact of the extraction system with the atmosphere (e.g., on sample changeovers) considerably affects the resolution of the membrane separator. These two problems were later overcome by these authors themselves (34) by using an APDC solution in an acetic-acetate buffer to broaden the linear range and thus avoid precipitation of the complexes in the aqueous medium, and by furnishing the system with a two-way valve, allowing air-free sample changeovers. The Swedish author applies his manifold to the determination of copper, nickel, lead, and zinc and achieves a sensitivity 15 to 20 times as high as that obtained by direct aspiration of aqueous samples, and a sampling frequency of 40 h^{-1} with as low as 0.3 mL min^{-1} MIBK consumption.

By use of an ingenious manifold involving injection of a complexed sample into a buffer carrier converging with a pure MIBK stream, Ogata et al. (35) determine zinc at concentrations between 20 and 800 ppb in biological and environmental samples, with a concentration factor of 2 and a relative standard deviation of 1.5%.

The metal extraction system introduced by Nord and Karlberg is incompatible with graphite furnaces, as the introduction into these of metals in the form of carbamate complexes is not recommended, owing to their high volatility. In 1984, Bäckström et al. (26) and Bengtsson and Johansson (22) simultaneously provided a solution to this problem based on application of a two-extraction method. The two methods involve the same chemistry and are based on the formation of hydrophobic complexes between metals and a mixture of dithiocarbamates which are extracted into trichlorotrifluoroethane (Freon 113). After separation, the organic phase is segmented by an acid solution of mercury(II), which forms stronger complexes with dithiocarbamates and thus displaces the metal of interest. In a second separation, the organic phase is discarded and the back-extracted metals contained in the aqueous phase are collected and detected by a graphite furnace/atomic absorption spectrometer.

The continuous flow extraction assemblies used are very similar, although Bäckström et al. achieve greater enhancement factors (between 25 and 30 without serious losses) by the use of higher flow rates of initial aqueous phase to back-extraction reagent.

A determination for lead in the urine of both exposed and unexposed adults by FIA-AAS/extraction has been reported (43). The liquid-liquid extraction is not carried out in a continuous fashion in an FIA system, but manually, a 20 µL portion of the extract subsequently being injected into a water stream leading into the nebulizer. The procedure is straightforward, fast, accurate, and reproducible, and affords recoveries between 96 and 105%. The results yielded are consistent with those obtained spectrophotometrically.

3.4 Indirect Determinations

Atomic absorption spectrometry can be used for the direct determination of 68 elements. Other elements cannot be determined directly because their resonance lines fall in the vacuum ultraviolet or because although they absorb in the accessible spectral region, it is extremely difficult to form and preserve a significant atomic population in both the flame and nonflame cells. Hence many authors have endeavored to broaden the scope of application of AAS by developing indirect determinations of elements not determinable in a direct manner (5,44,45).

The analytical potential of indirect determinations based on the formation of ion pairs and the use of FIA-AAS in conjunction with liquid-liquid extraction has been clearly shown in recent papers. Thus a modification of the FIA setup proposed by Nord and Karlberg has been used by our team to determine prechlorate traces in human serum and urine (27). The chemical foundation of the procedure is the formation of an ion pair between the cuproine-like chelate copper(I)-6-methylpicolinealdehyde azine and perchlorate which is extracted into MIBK. Perchlorate can thus be determined indirectly in the range 0.1 to 5 µg mL^{-1} at a sampling frequency of 45 h^{-1}, with a sensitivity threefold that achieved with with manual extraction and photometric detection.

Our team has also developed an automatic determination for anionic surfactants in wastewater (31) involving the formation of an ion pair between the detergent and a 1,10-phenanthroline-copper(II) chelate and its extraction into MIBK. A schematic diagram of the flow injection system used is shown in Fig. 11. The comprehensive study of variables potentially affecting the extraction efficiency and a comparison with the results obtained by the conventional technique allowed drawing the following conclusions: (a) chloride ion, indispensable to the manual procedure as a means of avoiding emulsion formation, has no effect on the extraction efficiency of the FIA

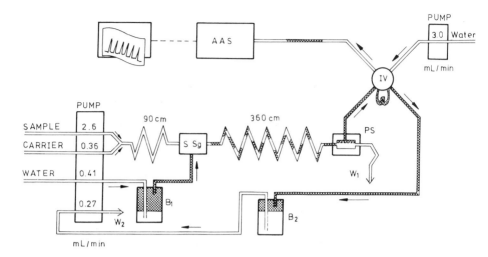

FIGURE 11   Flow diagram for the determination of anionic surfactants.   S Sg, solvent segmentor; PS, phase separator; IV, injection
valve; B_1 and B_2, displacement bottles for producing organic
streams; W_1 and W_2, waste.   [Reproduced with permission of the
American Chemical Society (31).]

procedure—the gentle mixing of the two phases in the extraction
coil probably prevents emulsion formation; (b) the FIA method features a wider pH range, greater sensitivity (0.1 mL μg^{-1} versus
0.06 mL μg^{-1} for the batch procedure), selectivity (see Table 3)
and precision (RSD 0.8%), and a higher sampling frequency (45 h^{-1});
and (c) the results are consistent with those found using methylene
blue, although somewhat more precise.

By use of an FIA system similar to that depicted in Fig. 11,
the authors have developed an indirect method of determination for
catonic surfactants in different types of water (P. Martínez-Jiménez,
M. Gallego, and M. Valcárcel, unpublished results) based on the
formation of an ion pair between the detergent and tetrathiocyana-
tocobaltate(II) and extraction into MIBK.   A sample containing between 0.3 and 13 μg mL^{-1} of the surfactant at pH 4 to 10 is continuously inserted into the system, where it converges with a carrier
solution of 0.04 M $Co(SCN)_4^{2-}$ in a 0.1 M acetic-acetate buffer of
pH 4.8.   The method yields acceptable extraction efficiency for flow
rate ratios in the range 9 to 14.   The detection limit for tetraheptyl-
ammonium bromide and dodecyltrimethylammonium bromide is 0.2 and
0.1 μg mL^{-1}, respectively.   Unlike its manual counterpart, the FIA
method requires no blank; in addition, it is more selective and

precise (1.2% RSD), and affords recoveries in cationic surfactants between 94 and 106% from both fresh water and wastewater.

Our team has shown the possibility of implementing sequential determinations with continuous liquid-liquid extraction with the individual (28) and sequential (29) determination of nitrate and nitrite based on the formation of an ion pair between these anions and the chelate bis(2,9-dimethyl-1,10-phenanthrolinate)copper(I) and extraction into MIBK. In a first step, the nitrite in the sample is oxidized to nitrate with Ce(IV) and total nitrogen is determined as nitrate. Then nitrite is converted to nitrogen with sulfamic acid, so that only the nitrate originally present in the sample is determined. The procedure allows the determination of nitrate-nitrite mixtures containing between 0.1 and 2.2 μg mL^{-1} of each ion, in ratios from 10:1 to 1:10 at a sampling frequency of about 20 h^{-1}. It has been applied successfully to the sequential determination of both anions in meat. The absence of a reducing column from the setup used avoids its cumbersome activation, which represents a significant advantage over other FIA procedures requiring the prior reduction of nitrate to nitrite.

4 ION EXCHANGE

Ion exchange is no doubt the separation technique most frequently used in flow injection atomic spectrometry. As prompted by Townshend (46), its widespread use could be the result of the ease with which ion-exchange minicolumns can be incorporated into FIA systems without detracting from the analytical properties of the system. One distinct advantage of ion exchange over other separation techniques is the possibility of changing the ionic composition of a solution without introducing undesirable interferents. This, together with its great simplicity, makes ion exchange a much favored separation technique.

The use of an ion-exchange column coupled to an atomic detector can be basically aimed at the removal of matrix and/or interferent effects, the preconcentration of metal traces, or the implementation of simultaneous determinations and/or speciation.

4.1 Basic Components of a Continuous Ion-Exchange System

The essential element of an on-line ion-exchange setup is a compact column where the exchange of ions between its solid packing material and a flowing solution takes place. The column is usually a glass, Teflon, Tygon, or PVC tube of one of the types commonly used in FIA, usually between 1.0 and 10.0 cm in length and 1.5 and 3.2 mm in inner diameter. Olsen et al. (47) use a FIAstar

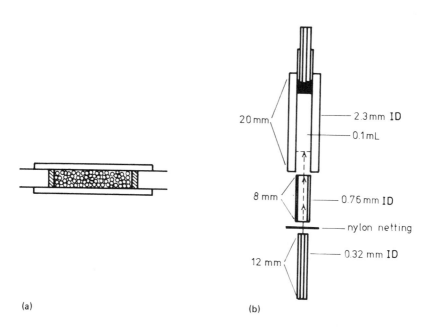

(a)   (b)

FIGURE 12   Construction of ion-exchange microcolumns: (a) column sealing; (b) column body. [Reproduced with permission of
the American Chemical Society (54).]

gradient tube (i.e., a Perspex block) into which a 5.0-cm-long
channel of 2.0 mm ID is drilled.

     The location of the minicolumn within the FIA system is as
varied as the features of the determination concerned can be. The
packing operation is started by stoppering one of the ends with a
small retainer, which can be a polyurethane foam piece (47,48), a
quartz (49–51) or dimethylchlorosilane-treated glass wool plug (52),
a 100-mesh nylon screen (53,54), or a polypropylene net (55).
Then the solumn is filled with an exchanger slurry. Once the column has been packed—avoiding as much as possible the introduction of air bubbles—it is sealed with a piece of the same material
used at the other end [Fig. 12(a)]. The procedure followed to
construct the column is described in detail in (54). As can be
seen from Fig. 12(b), the column body consists of a 20-mm piece
of Tygon tubing of 2.3 mm ID. The ends of the column are formed
by fitting a 12-mm piece of manifold tubing (0.32 mm ID) into another 8-mm piece of manifold tubing (0.76 mm ID) and are covered
with a piece of nylon netting tight enough to retain the resin particles in the column. After one column end has been sealed into
the column body with a drop of tetrahydrofuran, a slurry of

Chelex-100 in the hydrogen ion form is aspirated into the column
with a syringe. The other end is then sealed into the column body,
thereby obtaining a column with an overall volume of about 0.1 mL.

4.2 Types of Ion Exchangers

Anion exchangers are used in flow injection/atomic absorption spectrometry for interference removal purposes. Thus the FIA-AAS determination of calcium is interfed by anions such as sulfate and
phosphate (56), the disturbing effect of which can readily be overcome by using a strongly basic anion exchanger such as Amberlite
IRA-400 or De-Acite FF in chloride form. Most of the literature on
this topic deals with cationic exchangers intended for preconcentration of metal traces as a means of increasing the sensitivity of some
determinations.

Inorganic exchangers, despite their high exchange capacity
and rate and their great thermal and radioactive stability—usually
surpassing those of organic exchangers (57)—have scarcely been
used in flow injection analysis. Some authors use activated alumina
as an ion exchanger in FIA-ICP-AES, for both preconcentration and
matrix effect removal purposes.

The adsorptive properties of this oxide and its ionic exchange
capacity in chromatographic processes have been described thoroughly (58). Like most hydrous oxides, alumin has a stronger affinity
for polyvalent anions (particularly oxyanions) than for univalent
anions. McLeod et al (59-61) used activated alumina in its acidic
form for preconcentration and removal from the matrix of oxyanions
such as chromate, arsenate, borate, molybdate, phosphate, selenate,
and vanadate. As these polyvalent anions are strongly retained on
the resin, they are difficult to remove and usually call for the use
of strong bases, such as potassium hydroxide. Alumina has also
been used, though in basic form, for preconcentration of chromium-
(II) (62), using 2 M HNO_3 as eluent.

Chelating exchangers are the most commonly used in FIA-AAS
because of their great selectivity toward some transition metal ions.
The chelating agent incorporated into the resin should be chemical-
ly and mechanically stable, and feature a high exchange rate and
great exchange capacity (1 nmol or more of H^+ per gram). Most
of the chelating resins used so far are aminopolycarboxylic acids.
Such is the case with Dowex A-1 (48), Chelex-100 (47,49,54,63),
and Muromac A-1 (50,51). The most widely used exchanger, Che-
lex-100, is suitable for the selective sorption and separation of
heavy metals and some transition elements. The selectivity achievable depends on the pH of the sorption or eluting solution used.
The Chelex-100 resin is colorless, transparent, and readily swollen.
Hirata et al. (50) used a column packed with Muromac A-1 in acid

form for the preconcentration of chromium(III), titanium, vanadium, iron(III), and aluminum. This resin, brown and opaque, has a threefold higher concentration factor for chromium(III) at pH 3.8 than does Chelex-100 at the same pH, although both are iminodiacetate copolymers, due probably to their different particle size. Muromac A=1 is also purer than Chelex-100 and does not swell or shrink. It has also been employed for preconcentration of cadmium(II) (51).

Another weakly acid chelating resin used in FIA-AAS is made in Shangai, eastern China, by ICE. It has a phenol-formaldehyde base with salicylate functional groups and so far has been used to preconcentrate nickel, copper, lead, zinc, and cadmium in water (53,63,64), as well as in the simultaneous determination of free and complexed copper(II) (Cu-EDTA^{2-}) (65). A synthetic chelating resin prepared by modifying an ordinary anion exchanger with Cromoazurol S of remarkable stability to adsorption and regeneration has been used in the preconcentration of aluminium from hemodialysis fluids (66).

Another group of chelating ion exchangers contain functional groups immobilized on supports other than those used with styrene/divinyl benzene resins. The ready immobilization of 8-quinolinol on various supports has been exploited for the selective preconcentration of metal ion traces. The immobilization of this compound on silica gel supports (52) offers distinct advantages over its immobilization on organic polymers, as pointed out by Marshall and Mottola (52). First, silica is readily modified by a variety of silylating agents, allowing for a myriad of functional groups to be immobilized. Second, since the bound groups are at the support surface, high exchange rates are generally achieved, whereas some highly cross-linked organic polymer matrices may require hours for equilibration. Third, silica offers excellent swelling resistance, with changes in solvent composition having little effect on the support at pH 9. Therefore, silica surfaces are excellent choices for the immobilization of analytical reagents. On the other hand, the immobilization of 8-quinolinol on porous glass is more costly and results in a smaller breakthrough capacity than that achieved by immobilization on silica supports.

Finally, it is worth mentioning the advantageous use of a multidentate ligand, N,N,N'-tri(2-pyridylmethyl)ethylenediamine (Tri-PEN), immobilized on controlled-pore glass, in the determination of cadmium, copper, and zinc in complex matrices containing competing ions or soluble chelating ligands (55).

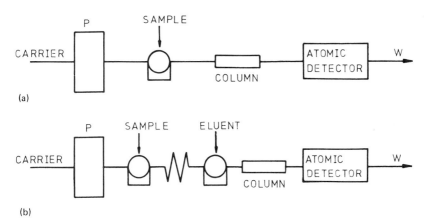

FIGURE 13 Single-line FIA-AAS systems with (a) one and (b) two injection valves. P, propelling system; W, waste.

4.3   Types of FIA-AS Ion-Exchange Configurations

Because of their variety, ion-exchange configurations used in FIA-AS are classified here according to the number of essential elements (column, propelling, and injection system) used and to their degree of complexity.

4.3.1   Systems with a Single Column

Single-column systems can be further subdivided into three groups:

1.   The simplest version uses a peristaltic pump and an injection valve [Fig. 13(a)]. The column is located immediately after the injection valve and there is a single channel through which the carrier is circulated. The valve functions to inject the sample (56) and, in most cases, to insert the eluant at a later stage (59-62). Some authors (64) use a multifunctional rotating sampling valve to carry out sequential sampling, ion exchange, elution, and AAS determinations. Kumamaru et al. (51) use a syringe instead of the injection valve to introduce the sample in a Teflon suction cup and then merge the sample with a buffer stream prior to passage through the column. The adsorbed ions are eluted by sucking 250 μL of 1 M HCl through the column, and the eluate is led directly to the ICP-atomic emission spectrometer.

2.   This group corresponds to configurations using a propelling system and two serial injection valves prior to the column. The propelling system can be either a gas pressure device (47,48, 66) or a peristaltic pump (52,65). As can be seen from Fig. 13(b), the first valve injects the sample into the carrier solution, which drives it to the minicolumn. The metals adsorbed on the exchanger

are subsequently eluted with a suitable eluant dispensed by the
second valve. During the elution operation, the first valve is
closed and its loop is filled with a fresh sample. Most of these
configurations include a coil between both valves, aimed to minimize the pump pulsations and/or to achieve complete homogenization
of sample and carrier.

These straightforward configurations may pose two problems,
arising from direct passage of the sample matrix to the detector and
from incomplete homogenization of sample and carrier. Moreover,
depending on the type of exchanger and eluant used, changes in
the degree of swelling of the resin resulting from the unidirectional flow may be encountered. The lack of sample homogenization can
be overcome by injection into a water stream and subsequent merging with an acid or alkaline buffer (47,52).

Malamas et al. (67) have devised a configuration of this type
for metal preconcentration and matrix isolation using two injection
valves and a peristaltic pump. The manifold employed is shown
ready for injection in Fig. 14. Either a water or a sample stream
is selected by valve 2 and directed through a six-way valve to a
microcolumn. While valve 3 is in this position, the auxiliary 400-μL
loop is filled with the eluant. When the valve is turned clockwise,
the flow forces the eluant in the loop through the column, so that
the retained metals are eluted and can be measured by AAS. A
variable-volume injection device is obtained by the time-controlled
operation of a pneumatic valve. Channel B, containing water, is
normally connected to channel C (the spectrophotometer inlet).

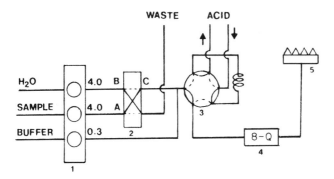

FIGURE 14 Schematic diagram of the flow injection trace enrichment
manifold. 1, Peristaltic pump; 2, timer-controlled, pneumatically
operated injection valve; 3, six-way rotary valve; 4, column with
240 μL of porous glass with immobilized 8-quinolinol; 5, flame. Flow
rates are given in mL min^{-1}. [Reproduced with permission of Elsevier Science Publishing (67).]

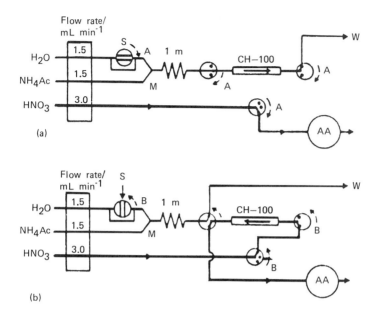

FIGURE 15 Two-line FIA-FAA system with directional valve and
peristaltic pump: (a) preconcentration cycle during which the sample volume injected by valve S, 2 mL, is merged with an ammonium
acetate buffer and then driven to the column; A, the corresponding
position of the valve for cycle (a), and (b) elution cycle, in which
the metal adsorbed on the Chelex-100 column is countercurrently
eluted by 2 M nitric acid and subsequently measured by the FAA
instrument; B, the corresponding position of the valve for cycle (b).
M denotes the merging point and the bold lines represent the lines
driving the liquids into the FAA instrument in cycles (a) and (b);
W is the waste. [Reproduced with permission of the Royal Society
of Chemistry (47).]

When valve 2 is actuated, channel A is linked to C for a preselected
time, and when it is deactivated, channel B is again linked to C.
This configuration has a number of advantages: (a) improved mixing results from direct merging of the sample with the buffer; (b)
the acid eluant flows through the nebulizer only during elution,
which lessens nebulizer corrosion; and (c) sensitivity is increased
as a result of using time and flow rate sampling rather than loop
sampling. Malamas et al. have applied a modification of this configuration involving higher sample and water flow rates (5.3 mL min^{-1})
and an acid loop volume of 670 μL to the preconcentration of metal
traces with a different ion exchanger (55).

The problem posed by swelling of resins such as Chelex-100 on passing from the initial NH_4^+ form to the hydrogen ion form after elution, resulting in increased capacity at the end of the column, has been tackled by use of a configuration also involving two injection valves but replacing the eluant with a solvent that allows upstream elution (Fig. 15). In this configuration (47) the sample is injected into a water stream and its pH is adjusted by merging with an ammonium acetate buffer. The direction of the flow is changed by means of a standard FIAstar valve which allows it to be directed through the Chelex-100 column in one direction during the preconcentration step and in the opposite direction during the subsequent elution step. Hence all directional functions, denoted by circles in Fig. 15, are executed by the valve. In the emptying position [A in Fig. 15(a)], the valve drives the eluant stream to the detector while the sample carrier and the nonretained fraction are wasted. In the filling position [B in Fig. 15(b)], the eluant flow traverses the microcolumn upstream, sweeping the retained cations out of the column into the nebulizer; simultaneously, the sample carrier is wasted and the injection valve is switched to the filling position. The only disadvantage of this configuration, the need for accurate timing of each cycle (which can be rather cumbersome if done manually), is outweighed by its advantages: (a) the eluant never reaches the nebulizer, so that the risk of blocking is avoided; (b) the sample is merged with the buffer, which results in improved mixing; and (c) upstream elution improves the column operation.

3. This group is integrated by configurations using two peristaltic pumps (Fig. 16). To ensure accurate cycle timing, Olsen et al. (47) designed a single-valve two-pump system including an electric timer capable of sequencing pumps 1 and 2 in a stop-go mode, each preconcentration-elution cycle being initiated by the turn of the injection valve. During the preconcentration cycle the sample is injected by switching the valve while pump 1 is in motion. Then the injected zone is mixed with the buffer in the coil and passed through the microcolumn. In the next sequence, pump 1 is stopped and pump 2 started, allowing the eluting acid to flow through the column in the opposite direction, thereby transporting the eluted metal to the instrument flame. When pump 1 is stopped, the liquids in the thus-closed circuit cannot move in either direction. The sampling cycle is completed when the peak appears, whereupon pump 2 may be stopped and pump 1 reactivated, thus establishing a high pH within the microcolumn and making it ready for the next preconcentration step. As in the system described above (Fig. 15), the seawater matrix never enters the spectrophotometer and the microcolumn is carefully regenerated prior to each sampling cycle while being operated in countercurrent fashion.

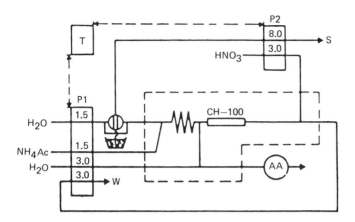

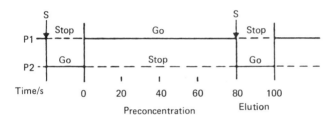

FIGURE 16 Fully automated FIA-FAA system operated via two peri-staltic pumps, the stop and go sequences of which are controlled by an electronic timer, T. The pumping rates are given in mL min^{-1} and sample volume is 2 mL. [Reproduced with permission of the Royal Society of Chemistry (47).]

Recently, Hirata et al. (49) proposed a manifold in which the sample is injected into a water stream flowing at a rate of 5 mL min^{-1} and later merged with a stream of 0.5 M ammonium acetate buffer prior to entering the microcolumn. The second peristaltic pump is used to introduce the eluant into the system. By means of two three-way stopcocks located in front of the column, one may sequentially switch between the preconcentration and the elution step. These authors (50) propose a very similar configuration for preconcentration of chromium(III), titanium(IV), vanadium(V), iron(III), and aluminum. In this assembly (Fig. 17), the sample is introduced in a continuous manner (sample loading times range between 30 and 180 s, depending on the sensitivity enhancement required) and merged with a buffer stream prior to passage through the microcolumn. Loaded samples are eluted for 20 or 25 s with

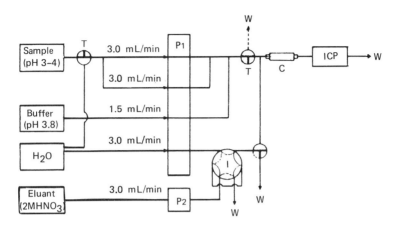

FIGURE 17   Scheme of manifold of on-line preconcentration FIA-ICP system. P_1 and P_2 , pumps; I, six-way valve; C, Muromac A-1 (length 3.2 mm, ID 8 mm); W, waste; T, three-way stopcock. [Reproduced with permission of the American Chemical Society (50).]

about 160 µL of 2 M HNO_3 at a pumping rate of 3.0 mL min⁻¹. A transient signal is obtained within 10 s after injection of the eluant. The Chelex-100 resin originally used by these authors was replaced with Muromac A-1 in their second configuration, thus avoiding swelling and shrinkage problems without detracting from the column resolution, despite the unidirectional nature of the flow.

4.3.2   Systems with Two Columns

Comparatively few FIA systems involving two columns have been reported so far.

Systems with One Peristaltic Pump and One Injection Valve. Fang et al. (53) designed the first flow injection manifold for metal preconcentration by ion exchange involving two parallel columns, which results in considerably improved efficiency. The key to their setup is a multifunctional three-layer sandwich-type rotary valve analogous to a commutation injector with three 3:2:3 commutation sections, two parallel sample loops, two eluant loops, and two ion-exchange columns arranged on the same rotor. By manually rotating the rotor at regular intervals, the sampling, ion-exchange, and elution operations are done sequentially and alternately on the two columns. As can be seen from Fig. 18, the sample (5 mL for an enhancement factor of 20 to 28), carrier, and eluant are pumped into their respective lines and the valve is switched manually at 90 s intervals. The various stages of the process taking place on the columns are described in the program listing in the figure.

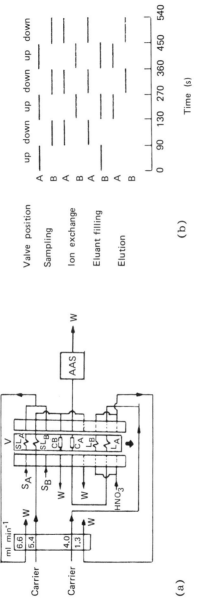

(a)

(b)

FIGURE 18 (a) Flow diagram of dual-column on-line preconcentration system. Subscripts A and B denote channels A and B, respectively. The multifunctional rotary valve, V, consists of S_A and S_B, samples; SL_A and SL_B, sample loops; C_A and C_B, ion-exchange columns; L_A and L_B, eluant loops; W, waste. Carrier: 0.1 M ammonium acetate, pH 5.5. (b) Program for the preconcentration process. [Reproduced with permission of Elsevier Science Publishing (53).]

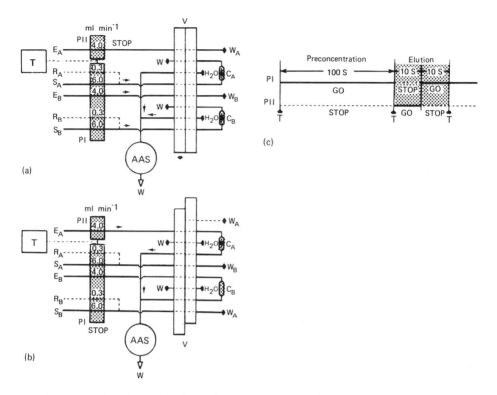

FIGURE 19 Manifold of dual-column on-line ion-exchange preconcentration system with flame atomic absorption detection. (a) Sampling
and preconcentration mode; (b) elution mode for column A. P_I and
P_{II}, pumps; T, timer; V, valve; S_A and S_B, samples; C_A and C_B,
ion-exchange columns; E_A and E_B, eluants for columns A and B;
W_A and W_B, waste lines for eluants A and B. (c) Time-sequenc-
ing program for valve operation. The points marked with T indicate
turn of the valve. The elution of both columns takes place while the
valve is in position (b), but with P_I and P_{II} sequenced stop-go and
go-stop, respectively. [Reproduced with permission of Elsevier
Science Publishing (63).]

The first peak of the analyte eluted from the column appears 185 s
after aspiration of the first sample, following two turns of the valve.
This manifold can be simplified merely by suppressing the loop holding the eluant and introducing this by direct aspiration through the
column into the nebulizer as prompted by the authors. Yet this involves the typical problems associated with nebulizer corrosion
through long exposure to acid eluants. The sole disadvantage of

this configuration lies in nonhomogeneous mixing resulting from direct injection of the sample in a buffer stream. There is little difference (only one of precision) in using concurrent or countercurrent elution.

Systems with Two Peristaltic Pumps. Fang et al. (63) have proposed a configuration aimed at improving in several respects the system described in the preceding section. The sample is merged with a buffer, time and flow rate sampling replace loop sampling, and samples are dealt with sequentially in parallel by the dual-column system. The configuration uses two peristaltic pumps whose stop-go intervals are controlled synchronously by a specially built timer triggered by a microswitch on the injector valve (a two-layer eight-way valve), which is actuated at each turn of the valve rotor. The manifold is set up as shown in Fig. 19. With the valve in the eluting position (b), samples A and B in the buffer are propeller by pump I through the pump tubes and valve to waste in order to flush out the previous sample. This is a preparation step preceding the sampling stage and should be done simultaneously with the elution of the analyte from the previous sample on column A. When the baseline is restored in the recording following elution of column A, the valve is switched manually to the sampling position (a), which actuates the microswitch, thereby starting the monitoring of the sampling time. At the end of the sampling period, pump I is stopped automatically. The valve is then switched back to the elution position (b), the second channel of the timer being actuated and pump II started as a result. The eluant is then pumped through column B, eluting the analyte directly into the spectrometer nebulizer, where it gives rise to a transient. By means of an ingenious modification of the valve, reported earlier by Jørgensen et al. (68), both the effluent emerging from the column during the preconcentration stage and the eluant containing no analyte are prevented from reaching the nebulizer. The timing program controlling the preconcentration and elution stages is listed to the right of the figure. This configuration affords an enahncement factor of 50 to 105, the highest achieved so far.

Another FIA setup used for preconcentration of various metals prior to their determination by ICP-AES uses two columns and as many valves and peristaltic pumps (54). One of the valves, an eight-way valve, is similar to that used in the system described above (Fig. 19). The scheme of the manifold is shown in Fig. 20. The injector system used features a number of advantages: (a) continuous aspiration of water to flush the nebulizer and stabilize the plasma; (b) countercurrent pumping of sample and eluant through the solumn to prevent resin packing; and (c) on-line control of the three operational modes available. Initially, the system

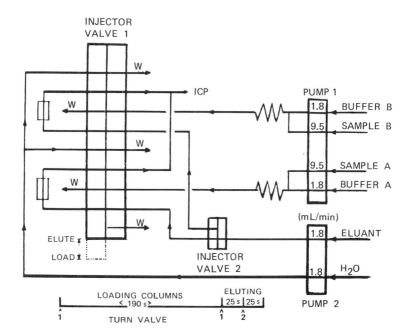

FIGURE 20 Scheme of injection manifold and flow system for on-line
FIA-ICP preconcentration system. [Reproduced with permission of
the American Chemical Society (54).]

operates in a conventional continuous aspirating mode by directly
aspirating standards and samples to the nebulizer through the line
carrying the water stream. The second and third operational modes
take advantage of the preconcentrating capabilities of either a single column or two parallel columns, respectively. In the two-column
mode, the sampling frequency is doubled as a result of simultaneously loading two separate samples into the two columns. The sample
loading times typically used range between 40 and 190 s, although
they can be increased according to the degree of preconcentration
required. The sample is merged with a buffer stream prior to entering the microcolumn. The eluant is pumped at a flow rate of
1.8 mL min^{-1} through a second injector, which directs it sequentially for one column to the other. A transient signal in the form
of an asymmetric peak is obtained within 5 s after injection of the
eluant, the baseline being restored in full after another 13 to 16 s.
Dual-column assemblies are generally more efficient and provide
higher sample throughputs that do single-column setups, although
special care must be taken to obtain two columns of identical
dimensions.

In the authors' opinion, future developments in the field of preconcentration manifolds should be aimed at achieving exchangers not liable to swell or shrink in passing from one to the other. As pointed out by Christian et al. (54), the scope of application of dual-column systems could be broadened by using a different packing material in each.

4.4 Theoretical and Practical Considerations

4.4.1 Features of Ion Exchangers

Ion exchangers feature a number of interesting physical (particle size, porosity, swelling, mechanical resistance) and chemical (exchange capacity and kinetics, selectivity) characteristics. Particle size is a factor of utmost relevance to ion-exchange separations. The smaller it is, the greater the separation efficiency will be. However, the concomitant increase in pressure drop across the column precludes the use of very small particles in FIA. Ion exchangers used in miniature columns for the preconcentration of metal traces are employed in a variety of sizes and pore diameters. Thus 8-quinolinol is used immobilized on controlled-pore glass (67) (50 nm pore size) or silica gel (52) in particle sizes ranging between 125 to 177 μm and 150 to 250 μm, respectively. The multidentate ligand TriPEN has also been used immobilized on controlled-pore glass (55) CPG-10, pore size 67 nm, particle diameter 75 to 125 μm). Activated alumina employed in FIA is usually 75 to 120 μm in particle size (59-62). The typical mesh size of various chelating resins also commonly used in FIA is 50 to 100 mesh for Chelex-100 (47,54,63) and Dowex A-1 (48), 60 to 100 mesh for Resin 122 (53,63-65), and 100 to 200 mesh for Muromac A-1 (50,51).

Ion exchangers can change in volume by absorbing or releasing solvent. Inorganic exchangers are scarcely prone to swelling. McLeod et al. (59-62) have found that activated alumina, whether used in acid or in basic form, does not undergo significant volume changes in the exchanging process. Silicate-based exchangers are also resistant to swelling because of their structural rigidity. Thus silica-immobilized 8-quinolinol provides excellent swelling resistance, to the decrement of the solution composition (52). Because of its excellent resistance to swelling or shrinking compared to polymer-based ion exchangers, porous glass (67) is another appropriate support for this reagent. Ion-exchange resins withstand swelling readily as a result of their elastic gel structure. The swelling properties of resins are not so important when using two columns manually (e.g., with the on-line low-pressure systems commonly employed in FIA). Fang et al. (63) have studied the swelling properties of Chelex-100 in depth. This resin is readily swollen, so that it can increase in volume by up to 100% in passing from its proton to its

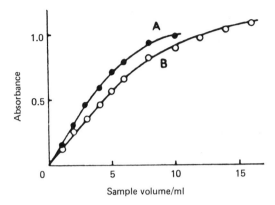

FIGURE 21 Breakthrough graphs for miniaturized column of Chelex-100 containing 25 mg of resin as obtained by injection of increasing volumes of standard solutions containing (A) 500 µg L^{-1} of lead and (B) 50 µg L^{-1} of cadmium. [Reproduced with permission of the Royal Society of Chemistry (47).]

NH_4^+ form. Volume changes result in major problems: erratic blockage, leakage, and flow rate oscillations. From their study these authors concluded that (a) the sampling period at a flow rate of 6 mL min^{-1} should not exceed 50 or 100 s for 0.5 and 0.05 M ammonium acetate, respectively, if the resin volume is not to change by more than 25% during a single operational cycle; (b) the column should be about three-fourths packed with resin in the H^+ form (washed with water) after conversion from the NH_4^+ form; (c) columns containing resins in the NH_4^+ form should never be washed with water; and (d) when not in use, columns should be converted to their H^+ form by normal elution with acid and washed thoroughly with water.

The mechanical resistance of resins has to do with the external wear of the exchanger particles through continuous use. For example, Chelex-100 wears significantly over time. In fact, Christian et al. (54) have found columns packed with this resin to have a lifetime of about 20 hr. The lifetime of Resin 22, by contrast, is well over 1000 cycles (53); however, if the column is left in 2 M nitric acid overnight, the exchanger may deteriorate to the extent of becoming virtually unusable. 8-Quinolinol is mechanically even more resistant. Columns containing this reagent immobilized on silica gel can be used for at least six months with no apparent deterioration in performance (52).

The breakthrough capacity of a resin is defined as the amount of metal ion that can be extracted per unit mass of solid under the operating conditions prevailing prior to being detected in the column effluent. It is influenced by a number of factors, including flow rate, temperature, column dimensions, particle size, and counterion concentration (52). Olsen et al. (47) determined the ion-exchange capacity of a miniature column containing about 25 mg of Chelex-100 by injecting increasing volumes of acidified standards of 500 µg if lead and 50 µg if cadmium per liter in seawater matrices into the system shown in Fig. 15. From the breakthrough graphs shown in Fig. 21, one may conclude that the breakthrough limit is attained at sample volumes between 4 and 5 mL. Further preconcentration (through increased sample volumes) thus calls for larger column sizes and amounts of resin. The exchange capacity varies from exchanger to exchanger; the capacity of Chromoazurol S is 0.5 nmol g^{-1} resin for aluminum (66), while that of 8-quinolinol immobilized on CPG is 5.5 µEq g^{-1} for zinc (67) and up to 37 µmol g^{-1} for copper when immobilized on silica gel (52). Breakthrough capacity generally increases with decreasing particle size; yet Mottola and Marshall use relatively large particles (150 to 150 µm) of 8-quinolinol immobilized on silica gel (52), as these result in improved column permeability when a low-pressure peristaltic pump is used at flow rates close to the nebulizer aspiration rate.

The time elapsed until the exchange equilibrium is reached is a major variable in ion-exchange separation techniques. In FIA, the adsorption of the analyte on the column and the rate of release from it have more pronounced effects than in a conventional ion-exchange preconcentration procedure, where the entire eluate containing a sample is collected before measurement. In conventional column operation, the analyte recovery is the same provided that no breakthrough occurs. In the FIA technique, the transient signal is obtained from an on-line elution. If we take into account the short residence time of a sample in a column within an FIA system (1 to 3 s), the kinetic behavior of each exchanger must be one of the factors determining its selectivity. Fang et al. (63) carried out a careful comparative study of the properties of three exchangers (Chelex-100, immobilized 8-quinolinol, and Resin 122). As shown in Table 4, the concentration efficiency varied considerably between elements and from exchanger to exchanger for a given element. Matrix effects aside, 8-quinolinol immobilized on CPG yielded the highest concentration factors. This was probably due to the faster exchange rate of its surface-bound chelating functional groups; the analyte is eluted more readily from the surface and is thus contained in a smaller eluate volume. The recoveries of the elements decreased roughly in decreasing order of their stability constants. Only lead was recovered satisfactorily; copper was poorly recovered, presumably because of the formation of its ammonium complex in the ammonium acetate buffer used (pH 9.5). Malamas et al. (67)

TABLE 4 Performance of Dual-Column Flow Injection System with Sequential Elution/Preconcentration and Different Chelating Resins

Resin	Characteristic concentration (μg mL^{-1})				Recovery[a] (%)				Concentration efficiency (fold min^{-1})			
	Cu	Zn	Pb	Cd	Cu	Zn	Pb	Cd	Cu	Zn	Pb	Cd
Chelex-100	0.5	0.12	1.5	0.2	99	92	95	93	88	50	70	60
8-Quinolinol	0.6	0.07	0.8	0.13	70	73	101	52	80	87	100	105
Resin 122 (weakly acidic)	0.5	0.13	1.0	0.2	97	94	101	44	88	47	83	60

[a]From a seawater matrix containing 3.1% NaCl, 1300 mg L^{-1} Mg, and 400 mg L^{-1} Ca. [Reproduced with permission of Elsevier Science Publishing (63).]

achieved better recoveries with 8-quinolinol for several elements, including copper(II) by using a phosphate buffer of pH 6.5. These authors studied in depth the competition of metal ions for the binding sites of the column. Silica supports endow 8-quinolinol with greater selectivity than CPG does (52). It is therefore not surprising that this has been the only exchanger used in the preconcentration of copper traces from water samples with matrices more complex than those of seawater.

The resin Chelex-100 provides good recoveries with all the metals studied by Fang et al. (see Table 4). The noninterference of alkaline metals can be attributed to the low stability constants of their imino-diacetate complexes (pK_1 -2.94 and -2.59 for magnesium and calcium, respectively) and to the high exchange capacity of the column. The results obtained with the weakly acidic Resin 122 are quite similar to those yielded by Chelex-100. According to the results obtained by these wuthors with a column packed with Resin 122, the recoveries of heavy metals are considerably better when using an ammonium acetate buffer of pH 9.5 (63) instead of pH 5.5 (53). From their comprehensive study, Fang et al. drew three interesting conclusions: (a) 8-quinolinol immobilized on CPG is the best performer whenever simple matrices (e.g., tap water or rainwater) are involved and maximum sensitivity is to be achieved; (b) Chelex-100 provides more accurate results with samples containing large concentrations of alkaline earth metals, although the sensitivity is reduced by 10 to 40%; and (c) Resin 122 is the best choice for samples with high alkaline earth metal contents, provided that analysis for cadmium is not required.

Obviously, the shortcoming of interference with the column exchanger remains to be solved. Thus, to the authors' minds, most of the exchangers proposed should be studied in greater depth with regard to their potential application to matrices more complex than seawater. Also, with the exception of the few papers commented on above, the kinetic aspects of exchangers, of paramount importance to this separation technique, have scarcely been deal with so far.

4.4.2 FIA Variables

Few of the papers reported to date on the topic of ion-exchange separations in FIA have covered the influence of the typical experimental FIA variables in the required depth. Increasing sample volumes is known to result in increasing preconcentration factors. Sample volumes in the range 0.2 to 100 mL have been used in FIA; yet they are typically between 1 and 5 mL. Increased volumes are affordable only when abundant samples are available (e.g., different types of water), although not with scarce samples such as biological fluids, which obviously do not allow for too high preconcentration factors.

The most influential FIA variable is no doubt the flow rate of the carrier driving the sample to the column or the eluate to the detector. When FIA is used in conjunction with AAS, flow rates are restricted to values greater than or equal to the aspiration rate of the nebulizer, resulting in too-large pressure drops when using packed columns. Fang et al. (53) have shown the eluant flow rate to be a critical factor. The results obtained by matching the eluant flow rate to the supposedly optimum aspiration rate of the nebulizer (5 mL min^{-1}) are far from satisfactory as a result of the time elapsed being insufficient for equilibrium to be attained during the elution. Lower flow rates give the analyte more time to desorb, so that it is present in greater concentrations in the eluate and yields stronger responses. If a low elution flow rate is used and the eluate is stored in a loop and subsequently flushed by a carrier stream at the instrument's recommended aspiration rate, the use of the theoretically optimum flow rate is found not to compensate for the decreased sensitivity resulting from the increased dispersion. Because of the different stabilities of the complexes formed in the elution process, each element has its own optimum elution flow rate.

Marshall and Mottola have studied the influence of the flow rate on the exchange capacity and found that breakthrough occurs earlier at high flow rates and increases with decreasing flow rate (52). In the preconcentrating FIA-ICP system reported by Christian et al. (54), the signal increases with increasing sample flow rate from 4 to 9.5 mL min^{-1} as a result of the increased sample volume. Yet the increasing trend is reverted at flow rates above 11 mL min^{-1} because of the decreased retention capacity of the column at flow rates higher than 9.5 mL min^{-1}, probably caused by a kinetic effect rather than by the column reaching breakthrough capacity. The typical flow rates used commonly range between 3.0 and 9.5 mL min^{-1} for the sample and are somewhat lower for the eluant.

4.4.3 Sensitivity

The use of ion exchange as an auxiliary technique with FIA-AS results indirectly in increased sensitivity with respect to the direct aspiration procedure. For a given system, the apparent sensitivity can be increased by increasing the sampling time or the injected sample volume [e.g., Malamas et al. achieve an enrichment factor of 500-fold from 100 mL of solution by using a sampling time of 25 min (67)]. The immediate result of using longer sampling times is obviously decreased sample throughput. Fang et al. (63) compared the concentration efficiency calculated by multiplying the concentration factor by the sampling frequency per minute provided by different systems in 1 min (sampling and elution times included). According to this criterion, the concentration efficiency of Malamas's system would be 20-fold per minute. Hence the results listed in

TABLE 5 Features of Ion-Exchange FIA/Atomic Spectroscopic Preconcentration Methods

Species	Detection	Preconcentration range	Sample volume (mL)	Sampling frequency (h^{-1})	RSD (%)	Type of exchanger	Applications	Ref.
Heavy metals	AAS	20-fold	1	30-60		Chelex-100	Seawater	47
Heavy metals	AAS	500-fold	100	2		8-Quinolinol[a]	Tap water	67
Heavy metals	AAS	20-28-fold	5	40	1.5-4.1	Resin 122	Water	53
Heavy metals	AAS	50-100-fold	10	60	1.2-3.2	Chelex-100 8-Quinolinol[a] Resin 122	Water	63
Heavy metals	AAS	5-450-fold	1-80	150-3		TriPEN[a]		55
Cu^{2+}	AAS	35-fold	10	15		8-Quinolinol[b]	Water	52
Ni^{2+}	AAS	20-fold			1.5	Resin 122	Water	66
Cu^{2+}, $(CuEDTA)^{2-}$	AAS		400 µL	90	1.0-1.6	Resin 122		65
Cu^{2+}, Mn^{2+}, Pb^{2+}	AAS		1	30	2	Dowex A-1	Soldering smoke	48
Al^{3+}	AAS	25-200 µg L^{-1} [c]	1	30	1-1.5	Synthetic resin (Chromoazurol S)	Hemodialysis fluids	66
Cd^{2+}	AAS	15-fold	4	24	<1.7	Chelex-100	Biological standard material	49
Heavy metals, Be^{2+} and Ba^{2+}	AES	10-30-fold		30	6	Chelex-100		54
Cr^{3+}, Cr^{6+}	AES	10-1000 µg L^{-1} [c]	0.2-2		2.2-1.1	Alumina	Reference water	59
Cr^{3+}	AES	10-1000 µg L^{-1} [c]	0.2-10		2.4	Alumina	Urine	62
Cd^{2+}	AES	25-fold	5	25	2.2	Muromac A-1	Biological material and waste water	51
Al^{3+}, Cr^{3+}, Fe^{3+}, Ti^{4+}, V^{5+}	AES	34-113-fold	3-18	17	2.3-4.4	Muromac A-1		50

[a] Immobilized on CPG.
[b] Immobilized on silica gel.
[c] Linear range.

Table 5 should also be compared from the point of view of the sampling frequency afforded in each case. The concentration efficiency
has also been found to vary with the particular element and exchanger (Chelex-100, 8-quinolinol, Resin 122) used (63) (see Table
4). For a given resin-element couple, the sensitivity provided by
the ICP-AES technique obviously excels that afforded by flame-AAS
(54).

4.5 Determination of Metals

The earliest semiautomated system exploiting the association ion-ex-
change/atomic absorption spectrometry was reported by Treit et al.
(69). Despite its continuous, unsegmented nature, it was not an
FIA system in the true sense, as it involved no sample injection.
It was applied to the determination of free copper in the presence
of EDTA and involved preparing several solutions containing identical concentrations of copper(II) in 0.100 M $NaNO_3$, to which increasing volumes of 0.02 M EDTA were added. To analyze for free
copper, the sample was inserted in a continuous fashion by a peristaltic pump into a minicolumn (5.0 cm length, 1.5 mm ID) packed
with Dowex 50W-X8 cation-exchange resin. The column was washed
with water for 2.5 min to remove interstitial solution and to establish a baseline AAS signal, and copper was then eluted with 0.02 M
EDTA for 1.5 min. The eluted copper was detected by an AAS peak
and the peak area measured. This cycle was repeated for each sample. A plot of free copper ion found versus amount of titrant added
provided the calibration curve. The flow system used by Treit et
al. included several rotary Teflon valves intended to direct the different flow lines. The method was optimized for the flow rate eluant composition, breakthrough, column length, and so on. It was
quite selective toward copper because the anionic titration product
formed by copper ($Cu\gamma^{2-}$) was not sorbed by the resin.

4.5.1 AAS Determinations

In Table 5 are summarized the features of the methods proposed so
far for the preconcentration and determination of various metals by
AAS. The earliest work in this field was reported in 1983 by Olsen
et al. (47), who published an interesting paper on the determination
of heavy metal traces (cadmium, copper, lead, and zinc) after concentration on a chelating resin (Chelex-100). These authors used
a variety of FIA setups of increasing complexity (see Figs. 15 and
16) aimed at solving the problems posed by their predecessors.
The carrier solution used was 0.05 M ammonium acetate of pH 7 to
10. The selectivity sequence found was $Cu^{2+} >> Pb^{2+} > Zn^{2+} >
Cd^{2+} > Ca^{2+} >> Na^+ >> H^+$. Hence the heavy metals assayed can
indeed be desorbed by strong acids. By injecting 180 µL of 2 M

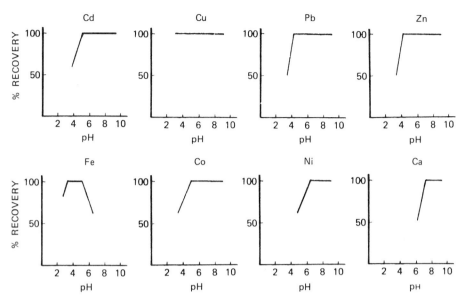

FIGURE 22 Metal uptake as a function of the pH. The metal standards were 0.1 (Cd and Cu) or 0.2 mg L^{-1}, and the sample volumes were 1.3 to 4.0 mL (sampling time 20 to 60 s). [Reproduced with permission of Elsevier Science Publishing (67).]

nitric acid, the metals initially contained in 1 mL of sample were found to be released into the much smaller volume of the acid zone for transport to the flame. The preconcentration cycle was quantitative provided that the carrier pH used was suited to the adsorption of the ions. The interference of seawater matrices was overcome by using a setup in which the column effluent was led to waste during the preconcentration step.

Malamas et al. (67) have demonstrated the possibility to augment the AAS sensitivity of cobalt, cadmium, copper, nickel, zinc, and lead indirectly by preconcentration on a column containing CPG-immobilized 8-quinolinol. The sample volume is controlled by pumping the sample directly through the column at a fixed low rate for a definite period of time. As shown by these authors, recoveries were quantitative at pH 6.5 for all the metals studied (see Fig. 22). As the pH decreases, protons displace the metal ions in the chelate, so that it is converted to neutral 8-quinolinol. Obviously, the metals forming the least stable chelates will be the most readily displaced by the protons. According to the authors, both samples and washing solutions should be buffered at high pH values to avoid the appearance of prepeaks as a result of pH changes.

Metals are eluted with 400 μL of a mixture of 1 M HCl and 0.1 M HNO$_3$. The sampling time and flow rate vary according to the enrichment factor achieved (500 for all metals except nickel, which amounted to 1000). The method was applied to the determination of the copper content in tap water, which is closely related to diarrhea in infants. The results obtained were consistent with those found by the graphite furnace technique, but more precise and readily gathered.

The preconcentration of the metals named above has also been tackled by Malamas et al. (55), with a similar configuration involving a minicolumn packed with a multidentate ligand immobilized on CPG as exchanger. These authors assayed N, N, N'-tetrakis(2-pyridylmethyl)ethylenediamine (TPEN) and N, N, N'-tri(2-pyridylmethyl)-ethylenediamine (TriPEN) as ligands and chose the latter, as it was found to excel over 8-quinolinol wherever soluble ligands were involved. The study of the pH dependence of the uptake (67) revealed that copper, silver, cadmium, cobalt, nickel, and zinc were taken up quantitatively, whereas lead, manganese, aluminum, calcium, and magnesium could not be retained quantitatively at any pH. The recovery of some metal ions, such as copper, silver, cadmium, nickel, and zinc, was found to decrease at high pH values owing to competing reactions (e.g., the formation of strong hydroxy complexes or precipitates). Quantitative elution required the use of acid solutions of 6 M HNO$_3$ or 6 M HCl, although a 1 M HNO$_3$ or 1 M HCl solution was sufficient if no copper or nickel preconcentration was needed or if quantitation was made from peak areas rather than peak heights, respectively. The experiments carried out on the recovery of cadmium traces in the presence of large amounts of calcium and magnesium showed the selectivity of the exchanger for transition metals over the alkaline earths.

The use of two parallel columns results in increased efficiency and sampling frequency, as it allows for sequential sampling, exchange, and elution. Fang et al. (53) use two columns packed with chelating Resin 122 for the preconcentration of heavy metals in water. Their system, shown in Fig. 18 and discussed in detail in Section 4.3, includes two sample loops and two eluant loops, as well as two minicolumns. The water sample (buffered with ammonium acetate of pH 5.5) and the sample-eluant carrier (2 M HNO$_3$) are introduced into their respective lines in the system. The volumes of the sample and eluant loops are 5 mL (for 20- to 30-fold concentrations) and 1.5 mL, respectively. Among the FIA variables studied by these authors, the eluant flow rate was found to be critical; yet it did not affect the precision of the method. To decrease the dispersion of the eluate plug in the carrier driving it to the detector, the valve was connected to the instrument via a piece (10 cm) of Teflon tubing. The authors chose a column length of 35 mm to ensure that back pressure in the system would be well within the limits withstood by a low-pressure peristaltic pump. The column diameter was found to have little effect on sensitivity in the range 2.2 to 3.5 mm. The flow rate selected for the exchange reaction, 5.4 mL min^{-1}, was apparently the highest possible not resulting in unacceptable breakthrough of the analytes because of insufficient interaction with chelating resin. Other features of the method are summarized in Table 5. Finally, the authors studied the potential interferents commonly found in water. Iron, manganese, zinc, nickel, copper, cadmium, and lead (2 mg L^{-1}) had no effect on the single metals quantified at concentrations in the range 10 to 200 μg L^{-1}.

Fang et al. (63) have reported another FIA configuration, using a dual-column ion-exchange valve for preconcentration of heavy metals. This system is more efficient in terms of sensitivity and time economy than those devised previously by these authors. Unlike the system described above (53), where one of the columns is eluted while the other is at the preconcentration stage (alternate operation), this novel dual-column setup allows processing of two samples simultaneously during ion exchange (the sampling process) and elution of the analytes sequentially by intermittent pumping with two pumps. This continuous introduction of sample and eluant into the column for a given period results in increased sampling frequencies. Obviously, both columns cannot be eluted simultaneously, as this would result in peak overlap. Thus one column is eluted while flow in the other is stopped, and after completion of the first elution, the process is reversed.

As stated above, 8-quinolinol immobilized on silica supports (SHQ) offers clear advantages over other organic polymer supports. These advantages have been exploited for the preconcentration of copper with a rather simple, het highly efficient FIA system (52) involving insertion of 10 mL of sample into a water stream later merging with a buffer (0.1 M ammonium acetate + HCl to pH 5.0) prior to entering a minicolumn containing 0.065 g of the exchanger. Although the preconcentration step takes 2.2 min, it is lengthened to 4 min to ensure complete extraction of the sample. Meanwhile a second valve is filled with eluant (1 M HCl/1.0 M HNO$_3$). Once the preconcentration has finished, the retained metal ions are eluted with 1.0 mL of the eluant and the eluate is carried directly to the nebulizer. By using the bath technique (0.2 g of SHQ was equilibrated with 10 mL of a standard metal ion solution for 1 h, after which the solid was removed by filtration and the metal ion concentration in the filtrate was determined), the content in each metal ion was calculated in terms of the pH to establish the selectivity of SHQ toward different metal ions. According to the authors, the individual selectivity of transition metals is not very good; however, the exchanger can be used for preconcentration and matrix isolation of aqueous samples containing large amounts of alkali and alkaline earth metal ions. The extraction of transition metals calls for relatively acid conditions.

Similar experiments carried out with nickel by the conventional procedure, using SHQ both alone and diazo coupled to (aminophenyl)-trimethoxisilane, yielded similar extraction-pH curves. The authors studied in depth the influence of the different variables. System dispersion occurs chiefly between the column and the flame, so that the length of the tube connecting the column to the instrument must be quite short. The concentrated eluant and dilute buffer used result in neutralization at the plug ends only, and hence in minimized dispersion, thereby allowing for a high local concentration of acid in the center of the plug for efficient elution of preconcentrated metals. The method features low detection limits and good reproducibility, and requires a minimum of sample manipulation. The method has been applied to the determination of copper in water.

Růžička et al. (65) have reported a system for simultaneous determination of free and EDTA-complexed copper ion by AAS, similar to that used by Cantwell et al. (69) for determination of free copper only, commented on above. Although the chemical foundation is the same for both systems, Růžička's is much simpler. In the first step, the sample containing both free and EDTA-complexed copper is injected into the carrier stream (water, 4 mL min^{-1}). The chelating ion-exchange column, located in front of the detector, separates the two components of the mixture. Free copper is retained on the column, while Cu-EDTA^{2-} ions flow through and are quantified by AAS. In the second step, the copper adsorbed on the column is eluted with 2 M HNO$_3$. Inasmuch as preconcentration occurs by elution into a smaller volume than the sample volume originally injected, and because the dispersion of the eluted zone is less than that for the complexed copper zone at the point of detection, the second peak, due to eluted copper, is narrower and therefore comparatively higher than the first. Of the three exchangers used—Chelex-100, 8-quinolinol immobilized on CPG, and Resin 122—the best results were provided by the latter, which competed much less markedly with EDTA for the copper ions than did the other two. The method provides relative standard deviations in the range 1.2 to 1.6% for Cu-EDTA^{2-} and 1.0 to 1.2% for Cu(II).

By use of a straightforward FIA setup, Hernández et al. (48) have developed a method for preconcentration and determination of manganese(II), lead, and copper(II). The manifold has two injection valves, which sequentially insert the sample and the eluant into the Dowex A-1 column. Although the injected sample and eluant volumes (1 mL and 180 to 200 µL, respectively) are roughly the same as those used by Olsen et al. (47) for the same system, this method is somewhat less sensitive, probably because of the different resin used. The authors apply the method to the joint determination of traces of the aforesaid metals in soldering smoke by differential elution with different solvents (0.3 M HCl for manganese, 1.0 M HCl for lead, and a 6.3 M ammonium-ammonia buffer for copper).

Hernández et al. (66) have also used this assembly for determination of aluminum in hemodialysis fluids. The serum samples are injected by means of a syringe through the chromatographic septum into the FIA system. A 0.5 M ammonium acetate buffer of pH 8 at a flow rate of 1.5 mL min^{-1} is used as carrier. After 2 min, required for the retention of aluminum, 100 μL of 1.0 M sodium hydroxide is injected to elute retained aluminum. Preconcentration and matrix effect cancellation are achieved simultaneously. The sensitivity of this technique can be increased by using the method of standard additions, which allows measurement of concentrations below 10 μg L^{-1} with a relative error of 2.4% for injected volumes of 1 mL without the need for prior mineralization.

One of the most recent applications of the preconcentration technique was reported by Hirata et al. (49) for the analysis of cadmium in biological standard reference material (Pepperbush, NIES No. 1). The manifold used includes two peristaltic pumps working independently in each preconcentration and elution cycle. The injected sample volume is 4 mL and the carrier used is water, later merged with a 0.5 M ammonium acetate stream. The eluant is 2 M HNO$_3$. A detection limit of 0.3 μg L^{-1} is achieved for cadmium. The method tolerates 1000-fold amounts of aluminum, copper, nickel, and zinc, and 3000-fold amounts of iron(II) and iron(III).

4.5.2 AES Determinations

Atomic emission spectroscopy has been used less often than AAS with ion-exchange or other separation techniques. Christian et al. (54) use a minicolumn packed with a Chelex-100 to increase the sensitivity of the determination of 11 elements by ICP-AES by use of the preconcentration multielement configuration depicted in Fig. 20. The sampling loading time is varied according to the required relative sensitivity. Of the 11 elements investigated, copper, zinc, cobalt, nickel, cadmium, berylium, barium, and manganese were found to yield the highest sensitivity at pH values between 5 and 9, while aluminum, iron, and lead reached maximum signal-to-noise ratios below pH 8. The decreased response yielded by these three elements above this pH is probably due to the formation of hydroxides not retained on the column. Nevertheless, the authors chose pH 9 for making simultaneous measurements of the 11 elements studied. The analytical signal was found to increase proportionally to the flow rate between 4 and 9.5 mL min^{-1}. The method is quite

TABLE 6   FIA/ICP-AES Deposition and Elution
Characteristics of Oxyanions on Activated Alumina

| Species | Deposition (%) | Elution (%) | |
		1 M NH₄OH	1 M KOH
Arsenate	>95	8	60
Borate	60	22	33
Chromate	>95	84	79
Molybdate	>95	84	84
Phosphate	>95	44	80
Selenate	>95	60	79
Vanadate	>95	41	58

Source: Reproduced with permission of the Royal
Society of Chemistry (61).

sensitive and reproducible (see Table 5) for all the elements assayed
except aluminum and iron.   The linear determinative range is 0.1 to
100 μg L^{-1}, where neither iron nor aluminum yield linear responses.
As the column efficiency is significantly less than 100% for aqueous
standards, the application of the system is limited to relatively
clean samples.

As stated above, inorganic exchangers have rarely been used
in FIA.   McLeod et al. use activated alumina as an ion-exchange
medium in FIA/ICP-AES for determination of phosphorus in steel
(60), speciation of chromium(III) and chromium(VI) in water (59),
determinations of oxyanions (61), and of chromium(III) in human
urine (62).

The speciation of chromium(III) and chromium(VI) is carried
out by injecting the sample into a carrier stream (0.01 M HNO$_3$ at
a flow rate of 1 mL min^{-1}).   On passing through the activated alumina (acidic form) column, the chromium(VI) in the sample carrier is
retained as chromium(III) flows through the detector.   A volume of
200 μL of 1 M ammonia is then injected to elute the retained chro-
mium(VI).   The recovery of this is calculated from the ratio of
peak areas obtained upon injection of 200 μL of 1.0 μg mL^{-1} chro-
mium(VI) with and without alumina column in the manifold.   The
flow rate of the acid carrier stream should not be so high as to
cause column deterioration.   Ammonium and potassium hydroxide
are equally effective in stripping chromium(VI) from the column.
The sensitivity of the method increases and its precision decreases
with increasing injected volumes.

This exchanger, also in acidic form, has been used for the preconcentration of oxyanions (60,61). The procedure involved is identical with that described above for the retention of chromium(VI). The carrier is also the same (0.01 M HNO_3), and both potassium of ammonium hydroxide at a 1 M concentration can be used as carriers. The deposition-elution features of the oxyanions on activated alumina are summarized in Table 6. According to their behavior, the different species can be classified into (a) those whose elution is less than 60% with 1 M potassium hydroxide (arsenate and vanadate); (b) those whose elution requires strong alkali, such as potassium hydroxide (selenate and phosphate); and (c) those to be eluted with strong or weak alkali (molybdate and chromate). The species classed as (a) can be desorbed by a strong eluant (e.g., 5 M KOH), which, however, rapidly ruins the column. This exchanger offers a number of advantages over other ion-exchange resins in the deposition and elution of oxyanions by FIA: namely, no significant changes over time, and long-term stability under both acidic and basic conditions.

These authors (62) have also used this exchanger in basic form to retain chromium(III), which is eluted with 200 µL of 2 M nitric acid. Prior to introduction of the next sample into the system, residual chromium is stripped from the column by injecting another portion of eluant. The alkalinity of the column is kept by using 0.01 M ammonium as carrier. Chromium(III) is quantitatively retained between pH 2 and 7. The method has been applied by its proponents to the determination of this element in human urine. The effect of the matrix was studied by injecting chromium(III) standards into synthetic urine samples, and enhanced emission was found in the matrix-matched solutions as a result of sodium and potassium not being retained on the column and calcium and magnesium being retained and subsequently eluted along with the chromium. The results obtained on urine standard reference material are quite acceptable, although low chromium(III) concentrations call for background correction.

Kumamaru et al. (51) have proposed an ICP-AES determination for cadmium in certified biological reference materials and wastewater samples. A column of iminodiacetate chelating resin (Muromac A-1) is used to preconcentrate the metal 25-fold for 5-mL samples. The sample is added to a Teflon suction cup from a pipette and elution is performed by aspiration with 1 M hydrochloric acid. Among the variables involved, the inner diameter and length of the column have some influence on the analytical signal obtained. The amount of eluant used is also influential, the greatest sensitivity being achieved for 0.25 mL of the acid. The emission line of cadmium (214.438) is not overlapped by that of platinum(IV) (214.423), as this is not retained on the column. The results obtained are

consistent with those found by the graphite furnace AAS technique
and with the certified values.

Muromac A-1 has also been used to increase the ICP-AES sensitivity to aluminum, chromium(III), iron(III), titanium, and vanadium (50). The sample (pH 3.8) is pumped through the column at
6.0 mL min^{-1}, mixed with a 0.5 M ammonium acetate buffer, and
sequentially eluted directly to the ICP nebulizer with a 2 M HNO$_3$
stream flowing at a rate of 3.0 mL min^{-1} along the manifold shown
in Fig. 17. The sensitivity to most of the elements studied increases
between pH 2.5 and 5. Above pH 5, the signal starts to decrease,
probably because of the formation of hydroxides. The sampling
time should be matched to the preconcentration factor required.
The analytical features vary from element to element. The signal
enhancement (fold increase per peak height) is high for chromium-
(III) and somewhat lower for aluminum, and the linearity range is
0.1 to 100 µg L^{-1} for all the elements except aluminum (2 to 100 µg
L^{-1}). Compared to Chelex-100, Muromac A-1 gives chromium(III)
concentration factors three times those afforded by the former. In
addition, this resin yields linear responses for aluminum and iron(III),
unlike the Chelex-100 resin used by Christian et al. (54). Hirata
et al. studied the potential interferences with the determination of
chromium(III) by use of Muromac A-1 and compared the results obtained to those provided by Chelex-100 in terms of tolerance to interferents. Elution of magnesium by 10, 100, and 1000 mg mL^{-1}
standards for 3-min column load times was 257, 1060, and 1407 mg
L^{-1} for the Muromac A-1, and 154, 586, and 763 mg L^{-1} for the
Chelex-100, respectively. Jointly eluted magnesium therefore interfered with the chromium(III) line (267.72 nm). Among metals, such
as manganese(II), iron(II), cobalt(II), nickel(II), copper(II), zinc-
(II), lead(II), aluminum(III), and iron(III), manganese showed positive spectral interference and aluminum negative interference with
the reaction between chromium(III) and iminodiacetate. Anions such
as nitrate, nitrite, carbonate, sulfate, and phosphate did not interfere at all. The authors concluded that the degree of interference
from foreign ions depends on both sample flow rate and column size.

ACKNOWLEDGMENTS

The authors wish to thank the CAYCIT for financial support received through project 2012/83, as well as M. Silva, P. Martínez-
Jiménez, and R. Montero, who helped in the team's work quoted in
this chapter.

REFERENCES

1. M. Valcárcel and M. D. Luque de Castro, Continuous separation techniques in flow injection analysis, *J. Chromatogr.* *393*:3 (1987).

2. J. Růžička and E. H. Hansen, *Flow Injection Analysis*, 2nd ed., Wiley, New York, 1988.

3. M. Valcárcel and M. D. Luque de Castro, *Flow Injection Analysis: Principles and Applications*, Ellis Horwood, Chichester, West Sussex, England, 1987.

4. M. Valcárcel, Selectivity and kinetics in analytical chemistry, *Analyst, 112*:729 (1987).

5. S. S. M. Hassan, *Organic Analysis Using Atomic Absorption Spectrometry*, Ellis Horwood, Chichester, West Sussex, England, 1984.

6. F. P. Treadwell, *Kurzes Lehrbuch der Analytischen Chemie*, Zurich, 1899.

7. J. K. Foreman and P. B. Stockwell, *Automated Chemical Analysis*, Ellis Horwood, Chichester, West Sussex, England, 1975.

8. J. K. Foreman and P. B. Stockwell, *Topics in Automatic Chemical Analysis*, Ellis Horwood, Chichester, West Sussex, England, 1979.

9. W. B. Furman (Ed.), *Continuous Flow Analysis. Theory and Practice*, Marcel Dekker, New York, 1976.

10. J. M. Skinner and A. C. Docherty, Determination of potassium in fertilisers by an automated ultra-violet absortiometric method, *Talanta, 14*:1393 (1967).

11. J. Minczewski, J. Chwastowska, and R. Dybczynski, *Separation and Preconcentration Methods in Inorganic Trace Analysis*, Ellis Horwood, Chichester, West Sussex, England, 1982.

12. A. Mizulke, *Enrichment Techniques for Inorganic Trace Analysis*, Springer-Verlag, Heidelberg, West Germany, 1983.

13. M. Valcárcel, M. Gallego, and P. Martínez-Jiménez. Indirect atomic absorption methods based on continuous precipitation in flow injection analysis, *Anal. Proc., 23*:233 (1986).

14. B. A. Petersson, Z. Fang, J. Růžička, and E. H. Hansen, Conversion techniques in flow injection analysis. Determination of sulphide by precipitation with cadmium ions and detection by atomic absorption spectrometry, *Anal. Chim. Acta, 184*:165 (1986).

15. P. Martínez-Jiménez, M. Gallego, and M. Valcárcel, Analytical potential of continuous precipitation in flow injection-atomic absorption configurations, *Anal. Chem., 59*:69 (1987).

16. P. Martínez-Jiménez, M. Gallego, and M. Valcárcel, Indirect atomic absorption determination of chloride by continuous pre-

cipitation of silver chloride in a flow injection system, *J. Anal. At. Spectrom.*, *2*:211 (1987).

16a. R. Montero, M. Gallego, and M. Valcárcel, Indirect atomic absorption spectrometric determination of sulphonamides in pharmaceutical preparations and urine by continuous precipitation, *J. Anal. At. Spectrom.*, *3*:725 (1988).

17. P. Martínez-Jiménez, M. Gallego, and M. Valcárcel, Indirect atomic absorption spectrometric determination of mixtures of chloride and iodide by precipitation in an unsegmented flow system, *Anal. Chim. Acta*, *193*:127 (1987).

18. R. Menaché, Routine micromethod for determination of oxalic acid in urine by atomic absorption spectrophotometry, *Clin. Chem.*, *20*:1444 (1974).

19. M. D. Luque de Castro and M. Valcárcel, Simultaneous determinations in flow injection analysis. A review, *Analyst*, *109*: 413 (1984).

20. P. Martínez-Jiménez, M. Gallego, and M. Valcárcel, Preconcentration and determination of traces of lead in water by continuous precipitation in an unsegmented-flow atomic absorption system, *Analyst*, *112*:1233 (1987).

21. L. Ebdom, *An Introduction to Atomic Absorption Spectroscopy*, Heyden, London, 1982.

22. M. Bengtsson and G. Johansson, Preconcentration and matrix isolation of heavy metals through a two-stage solvent extraction in a flow system, *Anal. Chim. Acta*, *158*:147 (1984).

23. J. A. Sweileh and F. F. Cantwell, Sample introduction by solvent extraction/flow injection to eliminate interferences in atomic absorption spectroscopy, *Anal. Chem.*, *57*:420 (1985).

24. T. M. Rossi, D. C. Shelly, and I. M. Warner, Optimization of a flow injection analysis system for multiple solvent extraction, *Anal. Chem.*, *54*:2056 (1982).

25. D. C. Shelly, T. M. Rossi, and I. M. Warner, Multiple solvent extraction system with flow injection technology, *Anal. Chem.*, *54*:87 (1982).

26. K. Bäckström, L. Danielsson, and L. Nord, Sample work-up for graphite furnace atomic-absorption spectrometry using continuous flow extraction, *Analyst*, *109*:323 (1984).

27. M. Gallego and M. Valcárcel, Indirect atomic absorption spectrometric determination of perchlorate by liquid-liquid extraction in a flow injection system, *Anal. Chim. Acta*, *169*:161 (1985).

28. M. Gallego, M. Silva, and M. Valcárcel, Determination of nitrate and nitrite by continuous liquid-liquid extraction with a flow-injection atomic-absorption detection system, *Fresenius Z. Anal. Chem.*, *323*:50 (1986).

29. M. Silva, M. Gallego, and M. Valcárcel, Sequential atomic ab-

sorption spectrometric determination of nitrate and nitrite in meats by liquid-liquid extraction in a flow-injection system, *Anal. Chim. Acta, 179:*341 (1986).

30. L. Nord and B. Karlberg, Extraction based on the flow-injection principle. Part 5. Assessment with a membrane phase separator for different organic solvents, *Anal. Chim. Acta, 118:*285 (1980).

31. M. Gallego, M. Silva, and M. Valcárcel, Indirect atomic absorption determination of anionic surfactants in wastewaters by flow injection continuous liquid-liquid extraction, *Anal. Chem., 58:* 2265 (1986).

32. L. Fossey and F. F. Cantwell, Characterization of solvent extraction/flow injection analysis with constant pressure pumping and determination of procyclidine hydrochloride in tablets, *Anal. Chem., 54:*1693 (1982).

33. L. Nord and B. Karlberg, An automated extraction system for flame atomic absorption spectrometry, *Anal. Chim. Acta, 125:* 199 (1981).

34. L. Nord and B. Karlberg, Sample preconcentration by continuous flow extraction with a flow injection atomic absorption detection system, *Anal. Chim. Acta, 145:*151 (1983).

35. K. Ogata, S. Tanabe, and T. Imanari, Flame atomic-absorption spectrophotometry coupled with solvent extraction-flow injection analysis, *Chem. Pharm. Bull., 31:*1419 (1983).

36. Y. Sahleström and B. Karlberg, Flow-injection extraction with a microvolume module based on integrated conduits, *Anal. Chim. Acta, 185:*259 (1986).

37. M. S. Cresser, *Solvent Extraction in Flame Spectroscopic Analysis,* Butterworth, London, 1978.

38. J. F. Tyson, C. E. Adeeyinwo, J. M. H. Appleton, S. R. Bysouth, A. B. Idris, and L. L. Sarkissian, Flow injection techniques of method development for flame atomic-absorption spectrometry, *Analyst, 110:* 487 (1985).

39. K. Fukamachi and N. Ishibashi, Flow injection-atomic absorption spectrometry with organic solvents, *Anal. Chim. Acta, 119:*383 (1980).

40. A. S. Attiyat and G. D. Christian, Nonaqueous solvent as carrier or sample solvent in flow injection analysis/atomic absorption spectrometry, *Anal. Chem., 56:*439 (1984).

41. N. Yoza, Y. Aoyagi, S. Ohashi, and A. Tateda, Flow injection system for atomic absorption spectrometry, *Anal. Chim. Acta, 111:*163 (1979).

42. L. Nord and B. Karlberg, Extraction based on the flow-injection principle. Part 6. Film formation and dispersion in liquid-liquid segmented flow extraction systems, *Anal. Chim. Acta, 164:*233 (1984).

43. J. L. Burguera, M. Burguera, L. La Cruz, and O. R. Naranjo, Determination of lead in the urine of exposed and unexposed adults by extraction and flow-injection/atomic absorption spectrometry, *Anal. Chim. Acta,* 186:273 (1986).

44. G. F. Kirkbright and H. N. Johnson, Application of indirect methods in analysis by atomic-absorption spectrometry, *Talanta,* 20:433 (1973).

45. M. Garcia-Vargas, M. Milla, and J. A. Perez-Bustamante, Atomic-absorption spectroscopy as a tool for the determination of inorganic anions and organic compounds, *Analyst,* 108:1417 (1983).

46. A. Townshend, Ion-exchange minicolumns, *Anal. Chim. Acta,* 180:49 (1986).

47. S. Olsen, L. C. R. Pessenda, J. Růžička, and E. H. Hansen, Combination of flow injection analysis with flame atomic-absorption spectrophotometry. Determination of trace amounts of heavy metals in polluted seawater, *Analyst,* 108:905 (1983).

48. P. Hernández, L. Hernández, J. Vicente, and M. T. Sevilla, Utilización de un sistema FIA-cambiador de iones y absorción atómica para la determinación de manganeso, plomo y cobre, *An. Quim.,* 81B:117 (1985).

49. S. Hirata, Y. Umezaki, and M. Ikeda, Determination of cadmium at the ppb level by column pre-concentration/atomic absorption spectrometry, *Bunseki Kagaku,* 35(2):106 (1986).

50. S. Hirata, Y. Umezaki, and M. Ikeda, Determination of chromium(III), titanium, vanadium, iron(III) and aluminum by inductively coupled plasma atomic emission spectrometry with an on-line preconcentrating ion-exchange column, *Anal. Chem.,* 58:2602 (1986).

51. T. Kumamaru, H. Matsuo, Y. Okamoto, and M. Ikeda, Sensitivity enhancement for inductively-coupled plasma atomic emission spectrometry of cadmium by suction-flow on line ion-exchange preconcentration, *Anal. Chim. Acta,* 181:271 (1986).

52. M. A. Marshall and H. A. Mottola, Performance studies under flow conditions of silica-immobilized 8-quinolinol and its application as a preconcentration tool in flow injection/atomic absorption determinations, *Anal. Chem.,* 57:729 (1985).

53. Z. Fang, S. Xu, and S. Zhang, The determination of trace amounts of heavy metals in waters by a flow-injection system including ion-exchange preconcentration and flame atomic absorption spectrometric detection, *Anal. Chim. Acta,* 164:41 (1984).

54. S. D. Hartenstein, J. Růžička, and G. D. Christian, Sensitivity enhancements for flow injection analysis-inductively coupled plasma atomic emission spectrometry using an on-line preconcentrating ion-exchange column, *Anal. Chem.,* 57:21 (1985).

55. M. Bengtsson, F. Malamas, A. Torstensson, O. Regnell, and G. Johansson, Trace metal ion preconcentration for flame atomic absorption by immobilized N,N,N'-tri(2-pyridylmethyl)ethylene diamine (TriPEN) chelate ion exchanger in a flow injection system, *Mikrochim. Acta, 3:*209 (1985).

56. O. F. Kamson and A. Townshend, Ion-exchange removal of some interferences on the determination of calcium by flow injection analysis and atomic absorption spectrometry, *Anal. Chim. Acta, 155:*253 (1983).

57. M. Marhol, *Ion Exchangers in Analytical Chemistry. Their Properties and Use in Inorganic Chemistry*, Elsevier, New York, 1982.

58. H. F. Walton (Ed.), *Ion-Exchange Chromatography*, Dowden, Hutchinson & Ross, Stroudsburg, PA, 1976.

59. A. G. Cox, I. G. Cook, and C. W. McLeod, Rapid sequential determination of chromium(III)-chromium(VI) by flow injection analysis-inductively coupled plasma atomic-emission spectrometry, *Analyst, 110:*331 (1985).

60. C. W. McLeod, I. G. Cook, P. J. Worsfold, J. E. Davies, and J. Queay, Analyte enrichment and matrix removal in flow-injection analysis-inductively coupled plasma atomic-emission spectrometry: Determination of phosphorus in steels, *Spectrochim. Acta Part B, 40:*57 (1985).

61. I. G. Cook, C. W. McLeod, and P. J. Worsfold, Use of activated alumina as a column packing material for adsorption of oxyanions in flow injection analysis with ICP-AES detection, *Anal. Proc. (London), 23:*5 (1986).

62. A. G. Cox and C. W. McLeod, Preconcentration and determination of trace chromium(III) by flow-injection/inductively-coupled plasma/atomic emission spectrometry, *Anal. Chim. Acta, 179:*487 (1986).

63. Z. Fang, J. Růžička, and E. H. Hansen, An efficient flow-injection system with on-line ion-exchange preconcentration for the determination of trace amounts of heavy metals by atomic absorption spectrometry, *Anal. Chim. Acta, 164:*23 (1984).

64. Z. Fang, S. Xu, and S. Zhang, Determination of trace amounts of nickel by on-line flow-injection analysis ion-exchange preconcentration and atomic-absorption spectrometry, *Fenxi Huaxue, 12(11):*997 (1984).

65. E. B. Milosavijevic, J. Růžička, and E. H. Hansen, Simultaneous determination of free and EDTA-complexed copper ions by flame atomic absorption spectrometry with an ion-exchange flow-injection system, *Anal. Chim. Acta, 169:*321 (1985).

66. P. Hernandez, L. Hernandez, and J. Losada, Determination of aluminium in hermodialysis fluids by a flow injection system

with preconcentration on a synthetic chelate-forming resin and flame atomic absorption spectrometry, *Fresenius Z. Anal. Chem.*, *325*:300 (1986).

67. F. Malamas, M. Begtsson, and G. Johansson, On-line trace metal enrichment and matrix isolation in atomic absorption spectrometry by a column containing immobilized 8-quinolinol in a flow-injection system, *Anal. Chim. Acta*, *160*:1 (1984).

68. F. J. Krug, B. F. Reis, and S. S. Jørgensen, *Proceedings of the Workshop on Locally Produced Laboratory Equipment for Chemical Education*, Copenhagen, Denmark, 1983, p. 121.

69. J. Treit, J. S. Nielsen, B. Kratochvil, and F. F. Cantwell, Semiautomated ion-exchange/atomic absorption system for free metal ion determinations, *Anal. Chem.*, *55*:1650 (1983).

6

Applications in Agricultural and Environmental Analysis

ELIAS A. G. ZAGATTO, FRANCISCO JOSÉ KRUG, and
HENRIQUE BERGAMIN, Jr. *Center for Nuclear Energy in
Agriculture, University of São Paulo, Piracicaba, São Paulo,
Brazil*

SØREN STORGAARD JØRGENSEN *Chemistry Deparmtent, Royal
Veterinary and Agricultural University, Frederiksberg,
Copenhagen, Denmark*

1 INTRODUCTION

The inception and initial development of flow injection analysis (FIA)
(1), which is based on the concept of nonsegmented continuous flow
analysis, was a consequence of increasing demand for chemical analysis in agriculture. Most of the early routine FIA systems, such
as those proposed in Brazil (2-15) and later in China (16-19), were
employed for determining chemical species in plant, soil, and water
samples. Also, the first FIA systems incorporating atomic absorption spectrometry (12) and inductively coupled argon plasma atomic
emission spectrometry (20) were proposed for plant analysis.

However, despite FIA's favorable characteristics of simplicity,
versatility, sampling rate, and sample and reagent consumption, a
relatively small number of FIA-AAS applications in agricultural analysis have been described (21). This is understandable, since conventional AAS is often employed successfully for agricultural analysis (22,23)—high sampling rates, good precision, and accuracy usually being achieved. Several elements (e.g., copper, zinc, manganese, and iron in plants) can be determined directly in the plant
digests.

Series of agricultural and environmental samples are often
characterized by a wide range of concentrations both within the
same element and from one element to another. Because of the
inherent narrow dynamic range of all absorption spectrometric methods (24), large and variable sample dilutions are frequently required
when batches of samples are analyzed by AAS for one or more elements. At the other end of the concentration scale, a number of
trace elements are present in such low concentrations that they cannot be determined reliably by flame AAS unless a preconcentration
procedure is carried out.

Therefore, sample conditioning prior to measurement is often
required. This step may include sample dilution and reagent addition, which limit precision and sampling rate when carried out manually. FIA-AAS systems are easily designed to overcome this disadvantage (12). The combination of FIA and AAS produces additional benefits, such as the possibility of simultaneous determinations (25), on-line sample preconcentration (19,26), dynamic range
expansion (27), improvement in system stability with consequent
achievement of better precision (28), automation of the standard
addition method (29), and interference characterization and suppression (30). Summarizing, the FIA-AAS combination causes a "syner-
getic effect" emphasizing the favorable characteristics of both flow
injection analysis and atomic spectrometry (31-36). The same
applies to FIA combined with flame photometry and ICP-AES.

In our experience, the combination of flow injection analysis
and atomic absorption/emission spectrometry is a simple and

convenient way of improving the precision and productivity of both routine and research laboratories in agricultural and environmental analysis.

2 POTENTIALITIES OF FIA-AS

2.1 Controlled Dilution

Sample dispersion in flow injection analysis depends on a combination of several factors which can be divided into two groups. The first group includes factors closely linked to the physicochemical properties of the involved solutions, such as diffusion coefficients, viscosities, and occurrence of chemical reactions (which depend on temperature and ionic strength). In practice, these factors are not easily modified. Parameters associated with system design, such as sample injected volume, flow rates, presence of confluent streams, characteristics of the analytical path, and so on, constitute another group of dispersion factors. The effects of these factors have been studied extensively, several mathematical models describing the FIA dispersion being available (37). However, in routine work carried out with simple FIA-AAS systems, the sample dispersion is usually empirically controlled by selecting the sample injected volume, the carrier-to-confluent streams flow rate ratio, the position of the confluence points, and the reactor length.

Changing the sample injected volume is the simplest way of changing dispersion and thus the sensitivity of any FIA procedure (38). In situations of very small injected volumes—large dispersion (38) in straight FIA systems—there is a proportionality between the volume injected and the recorded peak height. The width of the sample zone, hence the peak width, is almost unaffected by variations in the injected volume. The limit width, the peak width referred to zero injected volume (35), is approached. Therefore, the saturation index (35) tends to zero. In this situation, the concentration range for FIA-AAS systems is easily adjusted by suitable selection of the sample injected volume. For FIA-AAS systems employing measurements of peak areas, the injected volume may be modified without redefining the measurement time interval (20). If a larger integration time is required, the peak width may be increased by increasing the injected volume and simultaneously adding a confluent stream as far from the injection port as possible (35). The sensitivity cannot be increased at will simply by selecting larger sample volumes because as this factor increases, the peak width and the saturation index increase (35). Therefore, in situations of medium dispersion (38)—for straight FIA systems—increasing the injected volume increases both the height and the width of the recorded peak, the increase in peak width becoming more pronounced for larger

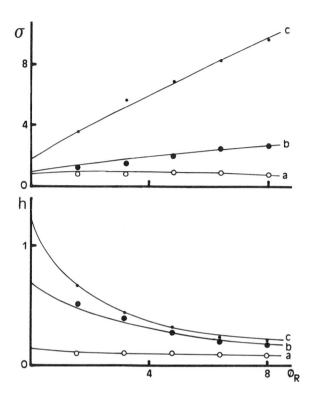

FIGURE 1   Effects of the confluent stream flow rate. ϕR, flow rate
of the confluent stream, in mL min^{-1}; σ, peak width at half-maximum,
in cm^3; h, peak height, in absorbance units.   Data are referred to
a spectrophotometric flow injection system into which a dye solution
is injected.   The flow rate of the sample carrier stream is 1.6 mL
min^{-1}, the confluence point is located 2 cm far from the commutator,
and the main reactor is 100 cm long.   Curves a, b, and c refer to
sample loops of 5, 50, and 200 cm, respectively.   For details, see
Ref. 35.   [Reproduced with permission of Elsevier Science Publishing (35).]

injected volumes.   Finally, in situations of limited dispersion (38),
the recorded peak height becomes almost unaffected by variations
of the injected volume, and a proportionality between injected volume and peak width is established (35).   The saturation index
tends to unity.   In practice, this situation should be avoided unless a very high integration time is required.   It should also be
emphasized that in situations where sensitivity is critical and a confluent stream is needed, the ratio between the sample carrier and
the confluent stream flow rates should be kept as high as possible.

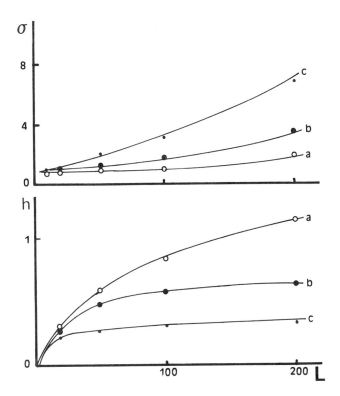

FIGURE 2 Effects of the length of the sample loop. This length, L, is expressed in cm. Curves a, b, and c refer to confluent stream flow rates of 0.0, 1.6, and 4.8 mL min^{-1}, respectively. Other symbols as in Fig. 1. For details, see Ref. 35. [Reproduced with permission of Elsevier Science Publishing (35).]

The effects of the confluent stream addition have been investigated (35) and are summarized in Figs. 1 and 2. At the confluence point, the sample zone undergoes dilution, being broadened simultaneously. The sample zone broadening causes a reduction in sample dispersion occurring downstream, which compensates for the dilution effect at the confluence point. When this compensation is almost total, a paradoxical situation is established in which the recorded peak height becomes practically unaffected by variations of the confluent stream flow rate. This phenomenon occurs in cases of small injected volumes, when the confluent stream is added near the injection point. Alternatively, when the postconfluence dispersion is negligible, the effects of the confluent stream addition are evident in the recorded peak. This occurs when large sample

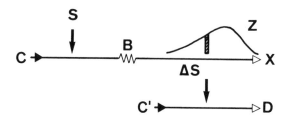

FIGURE 3   Zone sampling process. Following the sample injection,
S, into the carrier stream, C, and the dispersion inside the reactor,
B; a ΔS fraction of the dispersed zone, Z, is sampled and introduced
into the carrier stream, C', being directed toward detection, D. The
remainder of the zone, Z, directed toward X, can be either processed
or discarded. [Reproduced with permission of Elsevier Science Publishing (39).]

volumes are injected and/or when the confluent stream is added near
the detection unit.

    As a corollary, the position of the confluent stream addition
may also be an important dispersion factor, its relevance depending
on the injected volume (35). When the sample volume approaches
zero, the position of the confluence point is an important parameter
to be considered in the design of a FIA-AAS system. As the sample volume increases, the effect of this parameter on the sample
dispersion becomes less pronounced. With limited dispersion, the
position of the confluence point has no significant effect on the
sample dispersion.

    The effect of the reactor length on the recorded peak is ana-
logous to that of the damping factor (32). Increasing this length
decreases the peak height but simultaneously causes a peak broadening, which is often the limiting factor in sampling rate (37). Also,
the relevance of the reactor length as a dispersion factor diminishes
as this length increases. Therefore, the length of the reactor should
be selected as a compromise among sample dispersion, mixing conditions, and sampling rate.

    After defining the AAS operating conditions, the total flow rate
reaching the nebulizer, and the reactor length, the following guidelines may be useful:

1. If the sample concentration should be preserved, larger injected
    volumes should be employed, the confluent stream being added
    at low flow rate near the injection port.
2. If a large sample dilution is required, the injected volume should
    be as small as possible, the confluent stream being added at a
    high flow rate as far from the injection port as possible.

3. In both situations, the peak width should be compatible with the integration period.

In agriculture and related areas, high variability in the analyte concentrations are often found within batches of samples. This means a number of "out-of-range" samples to be reprocessed if a conventional FIA-AAS system is used. This limitation is overcome by using FIA systems that consider different regions of the dispersed sample, thus providing several analytical signals per sample, each associated with a given concentration range. These systems may include zone sampling (39), partially overlapping zones (40), or "gradient calibration" (41). In addition, time-based FIA-AAS systems (42) may be also employed for analysis of sample lots characterized by high variability in the analyte concentration.

2.2 Zone Sampling

The zone-sampling process was proposed by Reis and co-workers in 1981 (39) as an improvement in the determination of potassium in plant digests by flow injection atomic absorption spectrometry. In this process, a selected portion of a dispersed zone is sampled and introduced into another carrier stream (Fig. 3). This aliquoting process is very efficient in achieving a high degree of sample dispersion and is easily accomplished with a commutating device (34). The operation of a FIA system with zone sampling is described in detail in Section 3.2.

Zone sampling becomes attractive for the analysis of sample lots with high variability in the analyte concentration, as it permits different concentration ranges to be selected easily. This potentiality was exploited by Jacintho et al. (43), who determined calcium in waters, soils, and plants using the same FIA manifold, the required concentration ranges being adjusted by suitable selection of the t_S (34) values. Zone sampling is also useful to implement the standard addition method into a FIA system: only one standard solution may be employed, regardless of the number of the required additions and the addition levels (44). With programmable additions achieved by zone sampling and merging zones, the number of solutions with different concentrations to be used in the factorial experiments inherent to FIA design is reduced drastically. Also, this approach permits the achievement of a calibration curve by applying only one standard solution. The feasibility of zone sampling for simultaneous determinations was demonstrated when a FIA system was proposed for the determination of aluminum and iron in plant digests (25). The sampled aliquot was introduced into a manifold designed for the spectrophotometric determination of aluminum which required a high degree of sample dispersion. The remainder

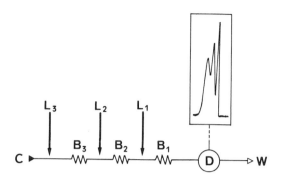

FIGURE 4 Partially overlapping zones. Several zones are established in the same carrier stream, C, partially overlapping peaks
being recorded. B_n, reactors; L_n, loops; D, flow through detector; W, waste. [Reproduced with permission of Elsevier Science
Publishing (34).]

of the sample zone was directed to the atomic absorption spectrometer, where iron was determined under conditions of limited dispersion (see also Section 3.5).

The zone-sampling process may also be exploited to achieve
images of the sample zone under real situations (45). In addition,
it should be mentioned that the sampler of a commercial flow injection analyzer is based on zone sampling (37).

2.3 Partially Overlapping Zones

In 1979, Mindegaard demonstrated that partial and reproducible
overlap between two injected zones could be achieved in flow injection systems (46). The system consisted of two merging carrier
streams, one transporting a sample zone and the other, a reagent
zone. The asynchronous and reproducible merging of these zones
allowed the analysis of very concentrated samples without prior dilution. Four years later, Hansen et al. (47) proposed a similar
FIA system for selectivity measurements. In this system, two different zones established into the same carrier stream overlapped
partially. Zagatto and co-workers (40) employed this approach to
record the overlapping peaks originating from two or three sequential sample zones. This approach, incorrectly termed "sequential
injection," is easily performed with commutation (34) (see also Section 3.6). The partial overlap between the sample zones is a consequence of the dispersion process occurring as they are transported toward detection. When the overlapping zones pass the detector, the measurement-time function is recorded (Fig. 4). Although any portion of the recorded function could be considered,

better precision is associated with the points of maxima and minima
of this function (39,40,48,49). Also, the absorbances related to
these portions are quantified without the need for sophisticated instrumentation. The process is, therefore, an alternative to FIA
techniques based on gradient exploitation.

With this approach, manganese in rocks was determined by
flow injection atomic absorption spectrometry (40), three concentration ranges being considered simultaneously (see also Section 3.6).
Similarly, partially overlapping zones were exploited to implement a
standard addition procedure in order to overcome matrix interferences in the determination of copper in ethanol by AAS (40). Recently, a FIA system employing a similar approach was proposed
for estimating and correcting for interferences in the determination
of chromium by flow injection atomic absorption spectrometry (30)
(see also Section 3.7).

In contrast to the zone-sampling approach, the use of partially
overlapping zones may be limited by the sample salinity. With zone
sampling, the most concentrated portions of the original sample zone
are not directed toward the spectrometer, while with the systems
employing partial overlap, these portions, although not being considered for analytical purposes, may cause depositions of solids on
the burner of the spectrometer. This limitation also holds for time-
based flow injection systems.

2.4 Time-Based Flow Injection Systems

In these systems, the peak width Δt at a preselected threshold level
constitutes the measurement basis. As the most concentrated portion
of the sample zone is not utilized for measurement, time-based systems do not require sophisticated detectors (42) and are characterized by a wider dynamical range than that of conventional FIA systems employing peak height measurement (27,32,42). When chemical
reactions are involved, these systems include the flow injection titrators (42,50,51). In time-based systems, discussed by Stewart (42),
linearity of the function Δt versus log C (C = analyte concentration)
should be attained. The linearity is improved by increasing the
sample dispersion, the damping factor of the AAS, and/or by lowering the threshold level (32).

As in systems with partially overlapping zones, high sample
salinity may cause deposition of solids on the burner of the spectrometer because for full exploitation of the time-based approach,
the most concentrated portion of the sample zone is introduced into
the nebulizer-burner for several seconds.

Most of the proposed time-based systems include atomic absorption spectrometry or spectrophotometric titrations. However, in
spite of its favorable analytical characteristics, little emphasis has

been given to the application of time-based FIA-AAS systems to
large-scale agricultural and environmental analysis (21).

2.5 Separations and Concentrations

Although several flow injection systems, involving dialysis (52),
distillation (53), gas diffusion (54), and so on, have been suggested,
most of the systems proposed for routine analysis of agricultural and
environmental samples employ ion-exchange or liquid-liquid extraction
as the separating and/or concentrating step.

Ion exchange was first employed in 1980 in connection with
flow injection analysis by Bergamin and co-workers (55), who proposed an automated procedure for the spectrophotometric determination of low levels of ammonium ion in natural waters. A microcolumn of Amberlite IRA 120 cation-exchange resin replaced the sample
loop of an electronically operated injector-commutator, so that in the
sampling position, a reproducible volume of sample passed through
the resin, allowing the adsorption of ammonium and other cations.
The sample was directed toward waste and not to the analytical path,
thus minimizing problems associated with matrix effects. Elution was
attained when the injector-commutator was switched to the injection
position, the ammonium ions being eluted by a sodium hydroxide solution acting as carrier stream. After elution, the Nessler reagent
was added, ammonium ion being measured colorimetrically on-line.
Although this system is still employed in CENA's laboratories, spectrophotometric FIA systems including ion-exchange columns have
seldom appeared in the literature (48,56,57). This could perhaps
be explained by recalling that the resin operation requires the
abrupt replacement of concentrated solutions, which causes the
"Schlieren effect" (8), often a limiting factor in spectrophotometry.
This limitation vanishes when AS is employed. In fact, the first
paper demonstrating the feasibility of ion exchange in FIA-AAS systems (26) does not mention the possibility of establishing undesirable concentration gradients, limiting the measurement. The increased
number of FIA-AS systems including resins (21,58-60) confirms that
the use of ion exchange in flow injection spectroscopy is promising.

Although concentration factors of up to nearly 100 have been
achieved in favorable cases (19,58,59), it should be emphasized that
the concentration efficiency depends considerably on the chemistry
involved in both the adsorption and elution steps. As a corollary,
either the ion exchanger or the species to be determined is usually
uncharged in the elution step. This favors the use of weakly acid
(including chelating) cation exchangers for metal determinations and
of strong alkali for eluting ammonium (55).

With regard to the placement of the resin column in the FIA
manifold, three principal configurations have been described (Figs.

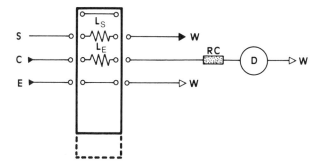

FIGURE 5   Flow diagram of a FIA system with a resin column placed in the analytical path.   S and E, sample and eluent solutions; L_S and L_E. loops; C, carrier stream; RC, resin column; D, flow-through detector; W, waste.   The boxed components are linked to the movable part of the commutator, the dashed line indicating the next commutation state.   The sites where pumping is applied are indicated by →.

5 to 7).   In the first configuration (Fig. 5), the resin column is placed in the analytical path.   The sample volume flowing through the ion exchanger is equal to the sample injected volume, the depleted sample flows through the detector, and elution is achieved after injecting the eluent solution.   In this system, concentration, elution, and resin conditioning occur during the flow injection period.   The resulting system can be manually operated.   This configuration, described first by Růžička and Hansen (38), has been also used in connection with conversion AAS methods (61,62).   In the configuration shown in Fig. 6, the sample loop of a loop-based injector is replaced by the resin column (55).   Higher concentration factors can be attained mainly because the sample volume flowing through the resin is not limited by the injected volume, being determined by the loading time and the sample aspiration rate. Therefore, electronic operation of the commutator is required. With this system, the depleted sample is not directed toward the detector.   Another advantage is that with this configuration, two resin columns can be employed simultaneously (19), with a consequent improvement of sensitivity and/or sampling rate.   With this approach, however, the resin cycle should be completed with only two steps, concentration and elution.   This drawback is circumvented by using the configuration shown in Fig. 7, a variation of this approach in which the initial sample volume is defined by a sample loop (26,59). Although electronic operation is no longer necessary, the sensitivity and/or sampling rate is decreased because emptying of a sample loop is involved.   With this approach, however, the resin cycle may

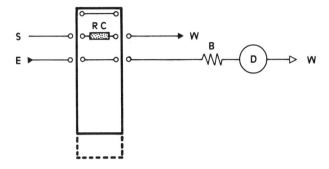

FIGURE 6 Flow diagram of a FIA system with a resin column replacing a sample loop/sample volume defined by sample aspiration
rate and sampling time. B, reactor. Other symbols as in Fig. 5.
[Reproduced with permission of Elsevier Science Publishing (55).]

include three steps: concentration, elution, and resin conditioning.
Finally, the feasibility of using a resin column as an ion sieve has
been demonstrated in spectrophotometry (56). This possibility does
not seem to have been investigated in combination with atomic spectroscopy.

Solvent extraction also has some attractions in flow injection
atomic spectroscopy. The establishment of a solvent segmented flow
prevents excessive sample dispersion (63). The nebulization is improved when an organic solvent is employed (22,23). However,
despite the favorable analytical characteristics, FIA-AAS systems
including solvent extraction have seldom been applied to large-scale
agricultural and environmental analysis (21,37,60).

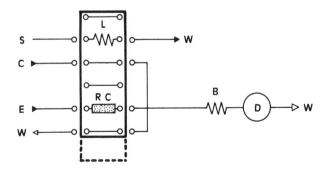

FIGURE 7 Flow diagram of a FIA system with a resin column replacing a sample loop/sample volume defined by loop. Symbols as
in Figs. 5 and 6.

3 SELECTED APPLICATIONS OF FIA-AS

The following procedures are employed in the laboratories of the first three authors for the routine analysis of large series of agricultural and environmental samples. Although other instruments can also be used, those listed below are used in connection with the selected procedures.

Constant head device: Ismatec model mp13 GJ4 peristaltic pump or
   similar provided with Tygon pumping tubes.
Injector-commutator: laboratory made. Consists of two external
   plates with a movable central bar (9) held together by two
   spring-loaded screws (Fig. 8). Holes (1 mm bore) are drilled
   through the commutator pieces, in accordance with the flow
   diagrams shown. Silicone rubber sheets with holes (about 0.5
   mm in diameter) are placed between the commutator plates to
   avoid leakage. A metal lever, manually or magnetically operated (39), is employed to move the commutator sliding bar.
   Alternatively, a B352 Micronal commutator is employed.
Manifolds: made of polyethylene tubes of the noncollapsible wall
   type (0.7 mm ID) which are wound around cylinders (2 cm
   outer diameter) in order to constitute the reactors. Tygon
   bushings (1.0 mm ID) are employed to connect the various portions of the manifold and also for insertion of the polyethylene
   tubes into the commutator plates. T-shaped Perspex connectors are also employed. A 3-cm-long narrow tube (about 0.3
   mm ID) interfaces the FIA manifold to the inlet of the spectrometer nebulizer.
Atomic absorption spectrometer: Perkin-Elmer model 503 atomic absorption spectrometer, operated according to the manufacturer's
   recommendations for maximum sensitivity with an air-acethylene
   flame. Unless otherwise stated, the damping factor is 1.2 s
   (TC2 setting on the instrument).
Flame photometer: model B262 Micronal instrument.
Recorder for AAS and flame photometry: Radiometer model REC61
   provided with a REA112 high-sensitivity unit. In the systems
   discussed in Sections 3.1 to 3.8, peak heights constitute the
   measurement basis.
Inductively coupled plasma atomic emission spectrometer: Jarrel-Ash
   model 975 AtomComp, operated according to the following conditions:
      Argon flow rates (1 min^{-1}): coolant 23, plasma 0, sample 0.5
      Observation height: 16 mm above load coil
      Power input to plasma (kW): incident 1.5, reflected <5

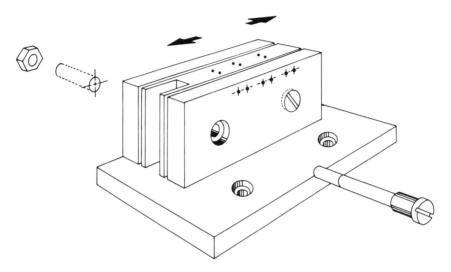

FIGURE 8 Manually operated commutator. A metallic lever is inserted in the central sliding portion. The length of this portion
is 5 cm.

> Wavelengths (nm): aluminum, 308.21; boron, 249.6; calcium,
>   317.9; copper, 324.7; iron, 259.9; mag-
>   nesium, 279.07; manganese, 257.6; phos-
>   phorus, 214.9; zinc, 213.8
> Integration period: 10 s

 The analytical signals are processed by a PDP-8E microcomputer, no alterations in the original software being required. For
each element, a two-point calibration is performed, a blank solution
being employed as the lower standard. Eventually, the ICP-AES
may also be connected to the Radiometer recorder.

 The procedures discussed in Sections 3.1 to 3.8 include the
sample preparation step, which is performed as follows:

 1. *Plant samples.* Dried and ground plant material with negligible fat content is solubilized by nitric and perchloric acids (6).
Weigh 0.750 g of plant material, transfer to a 75-mL digestion tube,
and add 7.5 mL of concentrated nitric acid. Mix well and leave at
room temperature for 1 to 2 h (or overnight). Then place tubes
into an electrically heated digestor block and set the temperature
at 160°C. Remove tubes from the block if the material starts to
rise in the tubes. When most of the nitric acid has evaporated
and a clear solution is obtained (about 15 min at 160°C), remove
tubes from the block and add 2 mL of concentrated perchloric acid.

Replace the tubes in the block and adjust the temperature to 210°C. The digestion is complete when a colorless solution is obtained and dense white fumes of $HClO_4 \cdot H_2O$ appear (about 15 min at 210°C). After cooling, the digests are diluted to 75 mL with distilled-deionized water. With this procedure, 1 mg L^{-1} of the metal determined in the final digest [approximately 0.25 M in perchloric acid (13)] corresponds to 100 mg kg^{-1} in the sample (dry basis). For plant samples with low metal contents, the final volume is completed only to 25 mL.

2. *Natural waters.* Natural water samples are collected into 1-L polyethylene bottles, preserved with the addition of 1 mL of concentrated nitric acid and analyzed within 24 h (64). Before being injected into the FIA systems, the aliquots are filtered through a 0.25 μm membrane filter.

3. *Rocks.* Rock samples (pyroxenite, iron ore, dunite, syenite, norite, granite, lujavrite, or similar) are decomposed in 40-mL Teflon bombs (40) following a procedure similar to that described by Langmyhr and Paus (65). The finely powdered samples (200 mg) are mixed with 1 mL of aqua regia and left at room temperature for at least 1 h. Then 4 mL of hydrofluoric acid is added; the bombs are closed and heated in a digestor block at 130°C for 2 h. After cooling, the bombs are opened, 1 mL of perchloric acid is added, and the open bombs are placed again in the blocks at 210°C. After about 1 h, when evolution of white fumes is observed, the bombs are cooled. If solids are still observable, the perchloric acid addition and evaporation should be repeated. The samples are then diluted to 20.0 mL with a 1% v/v nitric acid solution.

4. *Soils.* Soil extracts for the determination of exchangeable calcium, magnesium, sodium, and potassium are obtained according to a procedure similar to that recommended by Isaac and Kerber (66), which involves shaking 10.0 g of soil with 50 mL of 1 M ammonium acetate solution (pH 7). After filtration, the volume is completed to 100 mL with this solution.

3.1 Simultaneous Determination of Elements in Plant Digests by ICP-AES

Aluminum, boron, calcium, copper, iron, magnesium, manganese, phosphorus, and zinc are determined in plant digests. The standard solutions (10 mg L^{-1} boron, copper, iron, manganese, phosphorus, zinc; 20 mg L^{-1} aluminum; 100 mg L^{-1} calcium and magnesium) are also 0.25 M in perchloric acid. The FIA-ICP system of Fig. 9 is employed with the following parameters:

C: water carrier stream flowing at 3.2 mL min^{-1}
S: sample solution aspirated at 6 mL min^{-1}

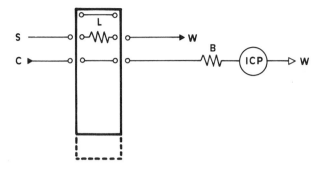

FIGURE 9 Flow diagram of the FIA-ICP system. ICP, spectrometer. Other symbols as in Figs. 5 and 6. [Reproduced with permission of Elsevier Science Publishing (20).]

L: 300-cm sample loop, inner volume about 1.5 mL
B: 40-cm-long tubular reactor

In the situation specified, the sample is aspirated to fill the
sample loop, its excess being discarded. When the commutator is
moved to the alternative position, the ICP spectrometer is initial-
ized (20) and the sample solution inside the loop is intercalated
into the carrier stream, being directed toward the spectrometer.
As the sample loop is much longer than the reactor, the D dispersion (38) approaches 1 and at the central portion of the sample
zone, a quasi-steady-state situation is established. The preburn
time of the system is adjusted so that only this portion of the sample zone in which the signal-to-noise ratio is maximal is quantified.
It is well known that when very large sample volumes are employed,
the sampling rate is sharply decreased (37,38). In this system,
this drawback is overcome by switching the injector commutator
back to the position specified in Fig. 9 immediately after the measurement, so that the last portion of the sample zone ("tail") still
inside the sample loop is pushed by the next sample toward waste
(34).

When this system was proposed (20), it was realized that the
inherent sensitivity of the spectrometric method was preserved and
that highly precise results (RSD similar to those related to normal
ICP operation) were obtained. The sampling rate of the original
procedure, about 20 samples per hour, was limited mainly by the
slow printout of the results. When a faster printer was employed,
the sampling rate was improved by a factor of about 2. Parallel
tests pointed out that the results are only slightly affected by variations in the sample salinity (or acidity) in the 0 to 1 M range.
This is probably due to the presence of the peristaltic pump, which

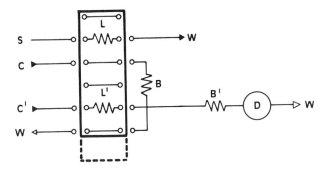

FIGURE 10 Flow diagram of the FIA system with zone sampling. C', second carrier stream; L', resampling loop; B', reactor. Other symbols as in Figs. 5 and 6.

provides a more constant flow rate entering the nebulizer than the standard pneumatic aspiration rate. During the use of this system for five years, adverse effects associated with the presence of perchloric acid in the samples have never been relevant. This is probably due to the reduction of the required sample volume and by efficient system washing. In any case, the possibility of reducing the perchloric acid concentration in the digests is recommended.

The system underwent some modifications as it was employed in the routine work. The sample loop was reduced to 100 cm and the carrier stream flow rate, to 1.6 mL min^{-1}; although the D dispersion was slightly increased, the resulting system was faster and the sample and argon consumption were reduced. Also, the possibility of adding an intermittent washing flow during the sampling period is promising (67).

3.2 Determination of Potassium in Plant Material by Flame Photometry with Zone Sampling

Potassium is determined in plant digests by flow injection flame photometry. The standard solutions, in the range 0 to 500 mg of potassium per liter, are also 0.25 M in perchloric acid. The FIA system of Fig. 10 is employed with the following parameters:

C and C': water carrier streams flowing at 6.8 mL min^{-1}
L and L': 20-cm sample loops (inner volume about 100 µL)
S: sample solution aspirated at 4 mL min^{-1}
B and B': 100-cm coiled tubular reactors

In the situation specified, the sample is aspirated to fill the first sample loop (L), which defines the injected volume, its excess being discarded. When the commutator is switched to the alternative

position, the sample selected volume is injected into the first carrier stream and the L' resampling loop is placed in the same path.
After a t_S time interval, when the dispersed zone is flowing through
the resampling loop, the commutator is moved back to the position
specified in Fig. 10. This movement causes the introduction of the
selected portion of the sample zone into the second carrier stream,
C', producing a second sample zone, which is similarly dispersed,
quantified, and discarded. The total sample dispersion depends on
the dispersions occurring into both carrier streams and on the Ts
value.

    This system is formed by two subsystems (48), which ideally
should be characterized by the same period (38). Therefore, although system dimensioning is not critical, this aspect should be
kept in mind when some parameter is modified. With a t_S value of
5 s, the central portion of the first dispersed sample zone is selected, better precision being attained (48). For sample lots with
potassium contents outside the foregoing range, other t_S values
should be defined experimentally.

    With this system, about 120 samples are analyzed per hour,
precise results being achieved (RSD usually less than 1%). It can
also be applied to plant digests obtained after a wet digestion employing sulfuric acid (68). In this situation the standard solutions,
in the 0 to 150 mg L^{-1} range, should also be 1.8 N in sulfuric acid,
and the sampling loops should be 50 cm long.

3.3 Practical FIA System Readily Dimensionable for Sodium, Potassium, Calcium, and Magnesium Determination in Natural Waters, Plants, Rocks, and Soil Extracts

Calcium, magnesium, sodium, and potassium are determined in natural waters, plant digests, soil extracts, and solubilized rocks by
automated atomic absorption or atomic emission spectrometry (69).
The standard solutions covering the concentration ranges specified
in Table 1 are also 0.1% v/v nitric acid (water analysis), 0.25 M
perchloric acid (plant analysis), 1 M ammonium acetate (soil analysis), or 1% v/v (rock analysis). The flow injection system of Fig.
11 is employed with B_1 = 10 cm, B_2 = 100 cm, and C = distilled-
deionized water. The other parameters are specified in Table 1.

    With the system of Fig. 11, only minor modifications such as
injected volumes, pumping tubes, and streamed solutions are required for the routine analysis of different matrices, in spite of
the high differences of the involved concentration ranges. For systems with large dispersion, the D value (38) undergoes a sixfold
increase when a 50-cm-long coil is used as the B_1 reactor (69). It
should be noted that for magnesium determination in soils and rocks,

TABLE 1   Parameters of the FIA System of Fig. 11[a]

Sample	Element	Concentration range (mg L^{-1})	ϕC (mL min^{-1})	R	ϕR (mL min^{-1})	Loop length (cm)
Water	Ca	0–10	5.8	b	1.0	250
	Mg	0–5	5.8	b	1.0	50
	Na	0–20	5.8	c	1.0	50
	K	0–10	5.8	d	1.0	250
Plant	Ca	0–100	5.8	b	1.0	e
	Mg	0–50	5.8	b	1.0	e
	Na	0–10	5.8	c	1.0	50
	K	0–500	5.8	f	1.0	e
Soil or rock	Ca	0–1000	1.0	g	5.8	e
	Mg	0–500	1.0	g	5.8	e
	Na	0–10	5.8	c	1.0	50
	K	0–10	5.8	d	1.0	250

[a]Ca and Mg are determined by atomic absorption spectrometry, and Na and K, by flame photometry.

[b]1.0% w/v La (as the nitrate or chloride salts) solution; also, 0.1 M nitric acid.

[c]250 mg K L^{-1} (as KCl) solution.

[d]250 mg Na L^{-1} (as NaCl) solution.

[e]The central part of the commutator is replaced, so that a trespassing bore of about 5 µL defines the injected volume.

[f]1000 mg Na L^{-1} solution.

[g]0.2% w/v La solution; also, 0.1 M nitric acid.

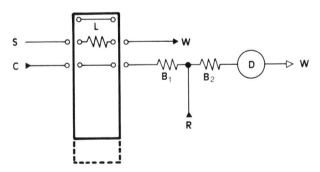

FIGURE 11   Flow diagram of the FIA system for calcium, magnesium, sodium, and potassium determination in natural waters, soil extracts, plants, and rocks. B_1 and B_2 reactors; R, confluent stream. Other symbols as in Fig. 5.

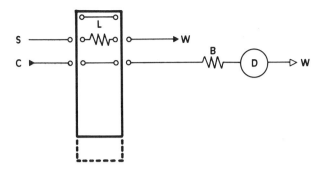

FIGURE 12 Flow diagram of the FIA system with limited dispersion.
Symbols as in Figs. 5 and 6.

the burner of the spectrometer should be placed obliquely (30°)
relatively to the light beam, to diminish the effective optical path.
In all the situations, sampling rates of 60 to 100 samples per hour
at the 1% carryover level are achieved with highly reproducible results (RSD usually less than 3%).

In routine work, the addition of the R solution has often been
accomplished by merging zones (10,12), as also discussed in Section
3.8. For some lots, strontium is used instead of lanthanum to overcome the iron, aluminum, phosphate, and silicate interferences,
quantitative interference suppression also being attained. In this
regard, the presence of calcium and magnesium impurities in the
strontium salt should be checked and the possibility of sulfate precipitation should be considered.

3.4 Simple FIA System with Limited Dispersion
       Applied to Plant Analysis

The FIA system of Fig. 12 is employed for the determination of copper, iron, zinc, and manganese in plant digests. As an example,
the determination of zinc is discussed.

The standard solutions (0.00 to 1.00 mg of zinc per liter are
also 0.25 M in perchloric acid. The system is dimensioned as follows:

C: water carrier stream flowing at 6.8 mL min^{-1}
S: sample solution aspirated at 6 mL min^{-1}
L: 200-cm sample loop, inner volume about 1.5 mL
B: 40-cm transmission line (as short as possible)

As in Section 3.1, the D dispersion (38) approaches unity and
the washing period is reduced by switching the commutator back to
the sampling position immediately after the recorded peak reaches

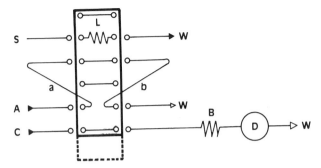

FIGURE 13 Flow diagram of the monosegmented FIA-AAS system. A, air; a and b, transmission lines. Other symbols as in Figs. 5 and 6.

its maximum. With a 300-cm-long sample loop, a damping factor of 4.5 s (TC3 setting on the spectrometer) can also be used, this modification being particularly useful when a noisy baseline is observed.

With the system of Fig. 12, about 140 samples are analyzed per hour at the 1% carryover level. If an electronically operated injector-commutator is available, carryover levels of up to 7% are acceptable, as the effect may easily be corrected by applying the simple equation

$$C_i = C_i^* - C_j K$$

where C_i and C_j are the concentrations of the ith and jth samples (the jth sample being injected immediately before the ith sample), C_i^* is the concentration calculated for the ith sample before carryover suppression, and K is the carryover level, in relative units, determined by means of the equation above, after injecting the most concentrated standard solution followed by the injection of a less concentrated standard solution.

With this approach, the sampling rate is improved by a factor of about 2. Sampling rate can also be improved by using monosegmented flow (70). In FIA monosegmented systems, the sample is injected between two air plugs, so that sample dispersion is reduced and some turbulence is established. Although monosegmentation was proposed as an alternative to achieve long sample residence times without excessive dispersion (70), it is also useful in FIA-AAS systems with limited dispersion. The cumbersome air remotion is not needed. The FIA-AAS monosegmented system of Fig. 13 operates as follows.

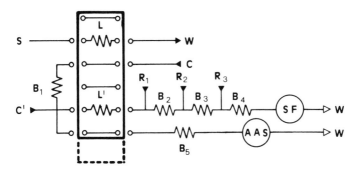

FIGURE 14   Flow diagram of the FIA system for the simultaneous determination of aluminum and iron. B_n, reactors; R_n, reagents; SF, spectrophotometer; AAS, atomic absorption spectrometer. Other symbols as in Figs. 5 and 10.   [Reproduced with permission of Elsevier Science Publishing (25).]

In the situation specified, the sample is aspirated to fill the sample loop and air is filling the a and b lines.   When the commutator is switched, the sample selected volume is placed between the a and b lines, the entire assembly being introduced into the main aqueous carrier stream.   For zinc determination in plants, the system is dimensioned as in Fig. 12, with the airflow rate of 3.9 mL min^{-1} and 10-cm-long a and b lines.

3.5   Simultaneous Determination of Aluminum and Iron in Plant Material

Aluminum and iron are determined in plant digests.   The mixed standard solutions, in the range 0.00 to 15.00 mg L^{-1} for both iron and aluminum, are also 0.25 M in perchloric acid.   The FIA system of Fig. 14, similar to that already proposed (25), is employed with the following parameters:

C and C': carrier streams of a 0.25 M hydrochloric acid solution, flowing at 6.0 and 3.2 mL min^{-1}

L and L': 100- and 40-cm sample loops (inner volumes about 500 and 200 μL)

S: sample solution aspirated at 4 mL min^{-1}

B_1, B_2, B_3, B_4, and B_5: 15-, 80-, 50-, 300-, and 40-cm reactors

R_1: 1% w/v ascorbic acid solution (prepared freshly), flowing at 3.2 mL min^{-1}

R_2: 0.02% w/v Eriochrome cyanine R solution, prepared as described earlier (13), flowing at 3.2 mL min^{-1}

R_3: 1.0 M sodium acetate solution, flowing at 3.2 mL min^{-1}

AAS and SF: atomic absorption spectrometer and spectrophotometer with a 178 OS Hellma flow cell (10 mm optical
path, 80 μL inner volume) set at 535 nm

In the situation specified, the sample is aspirated to fill the
first sample loop. When the commutator is operated, the initial sample volume is introduced into the first carrier stream being directed
toward the atomic absorption spectrometer, where iron is determined
under conditions of limited sample dispersion. After 8 s, when a
portion of the tailed part of the sample zone is still inside the resampling loop, the commutator is operated back to the position of
Fig. 14, producing a second sample zone which is pushed by its
carrier stream through the channel designed for the aluminum determination. Aluminum is determined under conditions of large dispersion (D about 40).

As zone sampling is concerned, the system requires electronic
operation. With this system, about 120 samples per hour are analyzed (240 determinations). For a typical sample, the precision of
the aluminum and iron determinations are about 1.5 and 2%, expressed as relative standard deviations. Stream splitting, not
recommended for situations involving rather different dispersions,
is avoided. Although the system is remarkably stable, recalibra-
tion after 60 samples is recommended. Other t_s values are employed
to expand the analytical range of the aluminum method. In the routine work, the approach has been extended for the analysis of other
chemical species.

3.6   Wide-Range Determination of Metals by Exploiting Partially Overlapping Zones

Flow injection systems with partially overlapping zones are employed
for the routine determination of elements occurring at wider ranges
in order to minimize the number of "out-of-range" samples. As an
example, the determination of manganese in rocks (40) is discussed.
The standard solutions, in the range 0.10 to 20.0 mg of manganese
per liter, are also 1% v/v in nitric acid. The FIA system of Fig.
15 is employed with the following parameters:

C: 1% v/v nitric acid solution flowing at 6.0 mL min^{-1}
S: sample solution aspirated at 5.0 mL min^{-1}
L_1 and L_2: 40- and 100-cm sample loops
B_1 and B_2: 100- and 200-cm coiled reactors

In the situation specified, the sample is aspirated to fill the
L_1 and L_2 loops, its excess being discarded. When the commutator
is switched, the loops are intercalated into the sample carrier stream
producing two zones initially separated by the B_1 delay coil. The

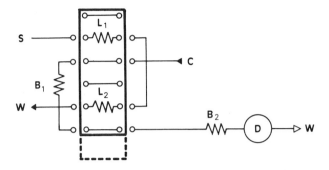

FIGURE 15   Flow diagram of the FIA system for wide-range determination of manganese in rocks—initial zone separation by a delay
coil.   Symbols as in Figs. 5 and 6.   [Reproduced with permission
of Elsevier Science Publishing (40).]

zones overlap while being transported toward detection, so that
two overlapping peaks are recorded.   The absorbance related to
the peak maxima and to the minimum between them ("valley") constitute the measurement basis, three calibration equations covering
three different concentration ranges then being achieved.

This FIA-AAS system, manually operated, is very stable; only
slight variations (usually 5%) in the coefficients of the calibration
equations are observed after an 8-h working period.   The typical
reproducibility of the FIA-AAS systems is maintained.   Similarities
between some results based on different calibration equations may
provide additional information on the accuracy.   The procedure is
characterized by a sampling rate of about 100 samples per hour at
the 0.1% carryover level.

Sampling rate is improved in about 30% when the system with the
configuration shown in Fig. 16 is employed.   The separation between
zones which determines the degree of overlap depends on the commutating times, so that the delay coil is not required.   With this system (L_1 = 20 cm), it is possible for the second peak to appear and
disappear during the wash period related to the first peak, so that
the implementation of the approach does not limit sampling rate.
In addition, the versatility of the system is improved.   Electronic
commutation is, however, required.

3.7   FIA System with Partially Overlapping Zones for Studying Interferences

Interference effects are investigated with a confluence FIA system
with partially overlapping zones.   As an example, the titanium and
iron interferences in the determination of chromium by AAS is

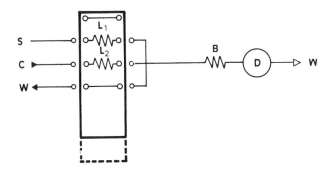

FIGURE 16 Flow diagram of the FIA system for wide-range determination of manganese in rocks—initial zones separation by time.
Symbols as in Figs. 5 and 6.

discussed (30). The system of Fig. 17 is employed with the following parameters:

C: 0.01 M hydrochloric acid solution flowing at 3.0 mL min^{-1}
S: injected solution aspirated at 3 mL min^{-1}
L_1 and L_2: 50- and 5-cm loops
B_1 and B_2: 50- and 100-cm reactors
Cr: 40 mg of chromium per liter standard solution, also 0.01 M hy-
     drochloric acid, flowing at 1.0 mL min^{-1}
R_1 and R_2: confluent streams flowing at 2.0 mL min^{-1}

For studying interference effects, different solutions are employed
as R_1 and R_2; the peak profiles of Fig. 18 refer to R_1 = 0.4 M

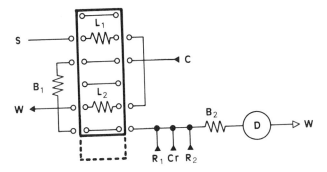

FIGURE 17 Flow diagram of the FIA system for investigating interference effects in the AAS of chromium. Cr, chromium standard
solution. Other symbols as in Figs. 5 and 14.

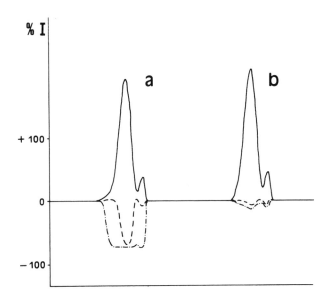

FIGURE 18   Recorder outputs obtained with the system of Fig. 17.
Situations a and b are emphasized in the text.   % I is the magnitude
of the interferences relative to the chromium steady-state absorbance.   The solutions injected as S are: ——, 40.0 mg chromium
per liter of standard solution; -----, 4000 mg titanium per liter of
standard solution; —·—·—, 4000 mg iron per liter of standard solution.

HNO_3 and R_2 = 0.4 M HCl (a) and to R_1 = 4.0 M HNO_3 and R_2 =
4% w/v oxine (CaH_7NO), also 0.4 M HCl solution (b).

The system operates as described in Section 3.6.   For interference investigations, a 50.0-mg chromium solution is injected as S,
two overlapping peaks on the chromium steady-state signal being recorded.   In this way, the concentration-time function is achieved, allowing the interferent concentration at any considered instant to be
estimated.   Thereafter, different solutions with a potential interfering species are injected, their effects on the chromium steady-state
absorbance easily being visualized (Fig. 18).

The recorder outputs shown in Fig. 18 clearly indicate that
both titanium and iron cause a depressive effect on the chromium
absorbance.   The magnitude of the titanium interference is roughly
proportional to the titanium concentration, whereas the iron interference tends to be independent of the iron concentration when this
concentration is higher than a threshold level.   In the latter situation, the addition of a constant iron concentration to all samples
and standards could be employed.   This is not required, however,
because both the titanium and iron interference effects are minimized
when a 4% w/v oxine solution is employed as the R_2 reagent (Fig. 18).

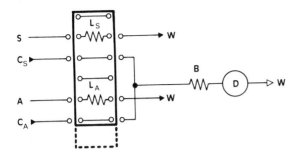

FIGURE 19 Flow diagram of a confluence FIA system with merging zones. A, solution to be added to the S solution; C_A and C_S, carrier streams; L_A and L_S, loops. Other symbols as in Figs. 5 and 6. [Reproduced with permission of Elsevier Science Publishing (34).]

The system permits the evaluation and characterization of interference effects and provides an efficient way for testing the procedures for interference suppression. All the investigations are carried out in the same conditions as in the system to be further proposed for large-scale analysis, so that the results are straightforward applicable.

3.8 FIA Systems with Programmable Additions

In atomic spectroscopy, the addition of a given solution is often required, especially when ionization suppressors or masking agents are required and/or the standard addition method is concerned. In routine work, this is often accomplished by using merging zones (10,12,34). A FIA system with merging zones (Fig. 19) operates as follows. After switching the commutator, the S and A volumes inside the L_S and L_a loops are introduced into their merging streams. The two established zones merge at the confluent point and the mixed zone is quantified. The merging-zone approach is also compatible with FIA straight systems (34). In this situation only small S and A volumes may be employed, the two zones are established into the same carrier stream, and the interaction between them is a consequence of the dispersion process.

By exploiting merging zones, a significant reduction in the amount of the A solution reaching the nebulizer-burner is attained (12). In addition, the procedure may be employed in connection with zone sampling and partially overlapping zones, as described in Sections 2.2 and 2.3.

REFERENCES

1. J. Růžička and E. H. Hansen, Flow injection analysis. Part I. A new concept of fast continuous flow analysis, *Anal. Chim. Acta, 78:*145 (1975).

2. J. Růžička and J. W. B. Stewart, Flow injection analysis. Part II. Ultrafast determination of phosphorus in plant material by continuous flow spectrophotometry, *Anal. Chim. Acta, 79:*79 (1975).

3. J. W. B. Stewart, J. Růžička, H. Bergamin F⁰., and E. A. G. Zagatto, Flow injection analysis. Part III. Comparison of continuous flow spectrophotometry and potentiometry for rapid determination of the total nitrogen content in plant digests, *Anal. Chim. Acta, 81:*371 (1976).

4. J. Růžička, J. W. B. Stewart, and E. A. G. Zagatto, Flow injection analysis. Part IV. Stream sample splitting and its application to the continuous spectrophotometric determination of chloride in brackish waters, *Anal. Chim. Acta, 81:*387 (1976).

5. J. W. B. Stewart and J. Růžička, Flow injection analysis. Part V. Simultaneous determination of nitrogen and phosphorus in acid digests of plant material with a single spectrophotometer, *Anal. Chim. Acta, 82:*137 (1976).

6. F. J. Krug, H. Bergamin F⁰., E. A. G. Zagatto, and S. S. Jørgensen, Rapid determination of sulphate in natural waters and plant digests by continuous flow injection turbidimetry. *Analyst, 102:*503 (1977).

7. L. Sodek, J. Růžička, and J. W. B. Stewart, Rapid determination of protein in plant material by flow injection spectrophotometry with trinitrobenzenesulfonic acid, *Anal. Chim. Acta, 97:*327 (1978).

8. H. Bergamin F⁰., B. F. Reis, and E. A. G. Zagatto, A new device for improving sensitivity and stabilization in flow injection analysis, *Anal. Chim. Acta, 97:*427 (1978).

9. H. Bergamin F⁰., J. X. Medeiros, B. F. Reis, and E. A. G. Zagatto, Solvent extraction in continuous flow injection analysis. Determination of molybdenum in plant material, *Anal. Chim. Acta, 101:*9 (1978).

10. H. Bergamin F⁰., E. A. G. Zagatto, F. J. Krug, and B. F. Reis, Merging zones in flow injection analysis. Part 1. Double proportional injector and reagent consumption, *Anal. Chim. Acta, 101:*17 (1978).

11. M. F. Giné, E. A. G. Zagatto, and H. Bergamin F⁰., Semi-automatic determination of manganese in natural waters and plant digests by flow injection analysis, *Analyst, 104:*371 (1979).

12. E. A. G. Zagatto, F. J. Krug, H. Bergamin F⁰., S. S. Jør-
     gensen, and B. F. Reis, Merging zones in flow injection ana-
     lysis. Part 2. Determination of calcium, magnesium and po-
     tassium in plant material by continuous flow injection atomic
     absorption and flame emission spectrometry, *Anal. Chim. Acta,*
     *104:*279 (1979).

13. B. F. Reis, H. Bergamin F⁰., E. A. G. Zagatto, and F. J.
     Krug, Merging zones in flow injection analysis. Part 3. Spec-
     trophotometric determination of aluminium in plant and soil
     materials with sequential addition of pulsed reagents, *Anal.*
     *Chim. Acta, 107:*309 (1979).

14. E. A. G. Zagatto, B. F. Reis, H. Bergamin F⁰., and F. J.
     Krug, Isothermal distillation in flow injection analysis. Deter-
     mination of total nitrogen in plant material, *Anal. Chim. Acta,*
     *109:*45 (1979).

15. E. H. Hansen and J. Růžička, FIA is already a routine tool
     in Brazil, *Trends Anal. Chem., 2:*5 (1983).

16. L. Sun, Z. Gao, L. Li, X. Yu, and Z. Fang, Determination
     of soil available phosphorus by flow injection analysis, *Fenxi*
     *Huaxue, 5:*586 (1981).

17. Z. Fang, Development of flow injection analysis (review),
     Turangxue Jinzhan, 10(2): 48 (1982).

18. Z. Fang and S. Xu, Determination of molybdenum at µg/L
     levels by catalytic spectrophotometric flow injection analysis,
     *Anal. Chim. Acta, 145:*143 (1983).

19. Z. Fang, S. Xu, and S. Zhang, The determination of trace
     amounts of heavy metals by a flow-injection system including
     ion-exchange preconcentration and flame atomic absorption
     spectrometric detection, *Anal. Chim. Acta, 164:*23 (1984).

20. A. O. Jacintho, E. A. G. Zagatto, H. Bergamin F⁰., F. J.
     Krug, B. F. Reis, R. E. Bruns, and B. R. Kowalski, Flow
     injection systems with inductively coupled argon plasma atomic
     emission spectrometry. Part I. Fundamental considerations,
     *Anal. Chim. Acta, 130:*243 (1981).

21. J. Růžička and E. H. Hansen, The first decade of flow injec-
     tion analysis: from serial assay to diagnostic tool, *Anal. Chim.*
     *Acta, 179:*1 (1986).

22. M. Pinta, *Spectrometrie d'Absorption Atomique,* 2nd ed., Mas-
     son, Paris, 696 p., 1980.

23. J. C. van Loon, *Analytical Atomic Absorption Spectroscopy—*
     Selected Methods, Academic Press, New York, 337 p., 1980.

24. S. R. Bysouth and J. F. Tyson, A comparison of curve fitting
     algorithms for flame atomic absorption spectrometry, *J. Anal.*
     *At. Spectrom., 1:*85 (1986).

25. E. A. G. Zagatto, A. O. Jacintho, L. C. R. Pessenda, F. J. Krug, B. F. Reis, and H. Bergamin F⁰., Merging zones in flow injection analysis. Part 5. Simultaneous determination of aluminium and iron in plant digests by a zone sampling approach, *Anal. Chim. Acta, 125:*37 (1981).

26. S. Olsen, L. C. R. Pessenda, J. Růžička, and E. H. Hansen, Combination of flow injection analysis with flame atomic-absorption spectrophotometry: determination of trace amounts of heavy metals in polluted seawater, *Analyst, 108:*905 (1983).

27. K. K. Stewart and A. G. Rosenfeld, Exponential dilution chambers for scale expansion in flow injection analysis, *Anal. Chem. 54:*2368 (1982).

28. W. R. Wolf and K. K. Stewart, Automated multiple flow injection analysis for flame atomic absorption spectrometry, *Anal. Chem., 51:*1201 (1979).

29. J. F. Tyson, J. M. H. Appleton, and A. B. Idris, Flow injection sample introduction methods for atomic absorption spectrometry, *Analyst, 108:*153 (1983).

30. M. C. M. Bezerra, Sistemas de injeção em fluxo para o estudo de interferências químicas na determinação de crômio por espectrometria de absorção atômica. Ph.D. thesis, Pontíficia Universidade Católica, Rio de Janeiro, 90 pp., 1987.

31. M. W. Brown and J. Růzicka, Parameters affecting sensitivity and precision in the combination of flow injection analysis with flame atomic-absorption spectrophotometry, *Analyst, 109:*1091 (1984).

32. O. Bahia F⁰., Quantificação temporal em sistemas FIA-AAS para a determinação de cobre em ligas metálicas, M.Sc. thesis, Universidade de São Paulo, São Carlos, Brazil, 81 pp., 1987.

33. J. Růžička and E. H. Hansen, Integrated microconduits for flow injection analysis, *Anal. Chim. Acta, 161:*1 (1984).

34. F. J. Krug, H. Bergamin F⁰., and E. A. G. Zagatto, Commutation in flow injection analysis, *Anal. Chim. Acta, 179:*103 (1986).

35. E. A. G. Zagatto, B. F. Reis, M. Martinelli, F. J. Krug, H. Bergamin F⁰., and M. F. Giné, Confluent streams in flow injection analysis, *Anal. Chim. Acta, 198:*153 (1987).

36. Z. Fang, S. Xu, X. Wang, and S. Zhang, Combination of flow-injection techniques with atomic spectrometry in agricultural and environmental analysis, *Anal. Chim. Acta, 179:*325 (1986).

37. M. Valcárcel and M. D. Luque de Castro, *Análisis por Inyección en Flujo*, Imprenta San Pablo, Córdoba, Spain, 450 pp., 1984.

38. J. Růžička and E. H. Hansen, *Flow Injection Analysis*, Wiley, New York, 207 pp., 1981.

39. B. F. Reis, A. O. Jacintho, J. Mortatti, F. J. Krug, E. A. G. Zagatto, H. Bergamin F⁰., and L. C. R. Pessenda, Zone sampling processes in flow injection analysis, *Anal. Chim. Acta, 125:*37 (1981).

40. E. A. G. Zagatto, M. F. Giné, E. A. N. Fernandez, B. F. Reis, and F. J. Krug, Sequential injections as an alternative to gradient exploitation, *Anal. Chim. Acta, 173:*289 (1985).

41. J. Růžička, Flow injection analysis—from test tube to integrated microconduits, *Anal. Chem., 55:*1040A (1983).

42. K. K. Stewart, Time-based flow injection analysis, *Anal. Chim. Acta, 179:*59 (1986).

43. A. O. Jacintho, E. A. G. Zagatto, B. F. Reis, L. C. R. Pessenda, and F. J. Krug, Merging zones in flow injection analysis. Part 6. Determination of calcium in natural waters, soil and plant materials with glyoxal bis(2-hydroxyanil), *Anal. Chim. Acta, 130:*361 (1981).

44. M. F. Giné, B. F. Reis, E. A. G. Zagatto, F. J. Krug, and A. O. Jacintho, A simple procedure for standard additions in flow injection analysis. Spectrophotometric determination of nitrate in plant extracts, *Anal. Chim. Acta, 155:*131 (1983).

45. E. A. G. Zagatto, O. Bahia F⁰., and H. Bergamin F⁰., Recording the real sample distribution and concentration/time functions in flow injection analysis, *Anal. Chim. Acta, 193:*309 (1987).

46. J. Mindegaard, Flow multi-injection analysis—a system for the analysis of highly concentrated samples without prior dilution, *Anal. Chim. Acta, 104:*185 (1979).

47. E. H. Hansen, J. Růžička, F. J. Krug, and E. A. G. Zagatto, Selectivity in flow injection analysis, *Anal. Chim. Acta, 148:* 111 (1983).

48. M. Martinelli, Estudos sôbre a determinação espectrofotométrica de cobalto em sistemas por injeção em fluxo empregando sal de Nitroso-R, M.Sc. thesis, ESALQ. Universidade de São Paulo, Piracicaba, Brazil, 78 pp., 1986.

49. M. Gisin and C. Thommen, Hydrodynamically limited precision of gradient techniques in flow injection analysis, *Anal. Chim. Acta, 179:*149 (1986).

50. J. Růžička, E. H. Hansen, and H. Mosbaek, Flow injection analysis. Part IX. A new approach to continuous flow titrations, *Anal. Chim. Acta, 92:*235 (1977).

51. A. U. Ramsing, J. Růžička, and E. H. Hansen, The principles and theory of high-speed titrations by flow injection analysis. *Anal. Chim. Acta, 129:*1 (1981).

52. J. Růžička and E. H. Hansen, Flow injection analysis. Part VI. The determination of phosphate and chloride in blood

serum by dialysis and sample dilution, *Anal. Chim. Acta, 87:* 353 (1976).

53. E. A. G. Zagatto, B. F. Reis, H. Bergamin F⁰., and F. J. Krug, Isothermal distillation in flow injection analysis. Determination of total nitrogen in plant material, *Anal. Chim. Acta, 109:*45 (1979).

54. H. Baadenhuijsen and H. E. H. Seuren-Jacobs, Determination of total CO_2 in plasma by automated flow injection analysis, *Clin. Chem., 25:*443 (1979).

55. H. Bergamin F⁰., B. F. Reis, A. O. Jacintho, and E. A. G. Zagatto, Ion exchange in flow injection analysis. Determination of ammonium ions at the µg/l level in natural waters with pulsed Nessler reagent, *Anal. Chim. Acta, 117:*81 (1980).

56. L. C. R. Pessenda, Determinação espectrofotométrica de molibdênio em digeridos vegetais com emprego de resina de troca iônica em sistema de injeção em fluxo, Ph.D. thesis, ESALQ, Universidade de Sao Paulo, Piracicaba, Brazil, 73 pp., 1987.

57. L. C. R. Pessenda, A. O. Jacintho, and F. J. Krug, Determinação espectrofotométrica de baixas concentracões de fosfato em águas naturais, através da preconcentracão em resina de troca iônica em sistema de injeção em fluxo, *Energ. Nucl. Agric., 5(2):*115 (1983).

58. S. D. Hartenstein, J. Růžička, and G. D. Christian, Sensitivity enhancements for flow injection analysis—inductively coupled plasma atomic emission spectrometry using an on-line preconcentration ion-exchange column, *Anal. Chem., 57:*115 (1983).

59. Z. Fang, J. Růžička, and E. H. Hansen, An efficient flow-injection system with on-line ion-exchange preconcentration for the determination of trace amounts of heavy metals by atomic absorption spectrometry, *Anal. Chim. Acta, 164:*41 (1984).

60. E. H. Hansen, Flow injection analysis, Ph.D. thesis, Technical University of Denmark, Lyngby, Denmark, 139 pp., 1986.

61. B. A. Petersson, Z. Fang, J. Růžička, and E. H. Hansen, Conversion techniques in flow injection analysis. Determination of sulphide by precipitation with cadmium ions and detection by atomic absorption spectrometry, *Anal. Chim. Acta, 184:*165 (1986).

62. A. T. Haj-Hussein, G. D. Christian, and J. Růžička, Determination of cyanide by atomic absorption using a flow injection conversion method, *Anal. Chem., 58:*38 (1986).

63. B. Karlberg and S. Thelander, Extraction based on the flow-injection principle. Part I. Description of the extraction system, *Anal. Chim. Acta, 98:*1 (1978).

64. American Public Health Association, American Water Works Association, and Water Pollution Control Federation, *Standard Methods*

for the Examination of Water and Wastewater, 14th ed., American Public Health Association, New York, p. 171, 1975.

65. F. J. Langmyhr and P. E. Paus, The analysis of inorganic siliceous materials by atomic absorption spectrophotometry and the hydrofluoric acid decomposition technique, *Anal. Chim. Acta, 43:*397 (1968).

66. R. A. Isaac and J. D. Kerber, Atomic absorption and flame emission photometry: techniques and uses in soil, plant, and water analysis, in *Instrumental Methods for Analysis of Soils and Plant Tissue*, L. M. Walsh (Ed.), Soil Science Society of America, Inc., Madison, Wis., 220 pp., 1971.

67. M. F. Giné, Análises de rochas por espectrometria de emissão atômica com plasma induzido empregando sistema de injeção em fluxo e método generalizado das adições padrão, Ph.D. thesis, ESALQ, Universidade de São Paulo, Piracicaba, Brazil, 95 pp., 1986.

68. J. A. Parkinson and S. E. Allen, A wet oxidation procedure suitable for the determination of nitrogen and mineral nutrients in biological material, *Commun. Soil Sci. Plant Anal., 6:*1 (1975).

69. F. J. Krug, E. A. N. Fernandez, I. A. Rufini, L. C. R. Pessenda, and A. O. Jacintho, Sistemas de injeção em fluxo para a determinação de Ca, Mg, Na e K em águas, plantas, rochas e sedimentos por espectrometria de absorção e emissão atômica, *Quim. Nova, 10:*165 (1987).

70. C. Pasquini and W. A. Oliveira, Monosegmented system for continuous flow analysis. Spectrophotometric determination of chromium(VI), ammonia and phosphorus, *Anal. Chem., 57:*2575 (1985).

7
Applications in Clinical Chemistry

ROY A. SHERWOOD and BERNARD F. ROCKS *Biochemistry Department, Royal Sussex County Hospital, Brighton, England*

1   THE USE OF ATOMIC ABSORPTION SPECTROMETRY
IN CLINICAL CHEMISTRY

The concentration of many of the metallic and semimetallic elements
in body fluids and tissues is of interest to workers in several of
the fields that constitute clinical chemistry, including toxicology,
therapeutic drug monitoring, and nutrition. Since its introduction
in the mid-1950s, atomic absorption spectrometry has become the
accepted method for the measurement of these elements. The technique has been refined steadily over the three decades since its
inception enabling the measurement of elements at progressively
lower concentrations, thus permitting the determination of elements
present only in ultratrace amounts in the human body.

The majority of measurements are performed on blood specimens,
although urine, tissue, hair, and other body fluids are occasionally
assayed, a typical example being the measurement of copper or iron
in samples of liver tissue obtained by biopsy. The distribution of
the various elements in blood varies enormously, but in most cases
the sample is separated into its two main components, the cell fraction and the serum or plasma. Serum is obtained by allowing the
blood to clot in a glass tube, and sufficient time must be allowed
for the clot to form and retract adequately. Serum is preferable
to work with as there is less protein precipitation, and the possibility of contamination from the anticoagulant required for the preparation of plasma is eliminated. Serum is, however, obtained in
poorer yield than plasma, and also because plasma separation is
quicker, it is used in many laboratories. Some elements, however,
are contained primarily in the cells (e.g., lead, cadmium, and selenium) and whole blood measurements are commonly used for these
elements.

Contamination is a major problem in the measurement of many
of the metallic elements. The anticoagulant used must be selected
with care, as it may introduce contamination; for example, lithium
heparin, a commonly used anticoagulant, is obviously unsuitable for
specimens on which lithium measurements are to be performed.
Other less obvious sources of contamination include the syringes
and needles used to obtain the specimen and the tubes or bottles
in which the samples are stored. This is particularly important if
the trace elements are to be determined. The foregoing difficulties
apply equally well to the use of body fluids other than blood, and
other sample media, hair, for example (1), present their own problems as to contamination from external sources.

The elements most commonly measured by atomic absorption
spectrometry (AAS) in the hospital laboratory are detailed in Table
1. Table 1 also shows the type of sample, a typical "normal range,"
and representative disease states or conditions in which measurement

TABLE 1 Elements Commonly Measured in the Clinical Laboratory Using AAS

Element	Sample medium	Approximate normal range	Common conditions in which the element is measured
Aluminium (Al)	Serum	<2 μmol L^{-1}	Al toxicity in hemo-dialysis
Cadmium (Cd)	Blood	<130 μmol L^{-1}	Cd toxicity
Calcium (Ca)	Serum	2.2-2.6 mmol L^{-1}	Bone disease
Cobalt (Co)	Serum	8.5-136 nmol L^{-1}	Cobalt toxicity
Copper (Cu)	Serum	12-26 μmol L^{-1}	Wilson's disease
	Urine	0.2-1.0 μmol L^{-1}	Menke's disease
Iron (Fe)	Serum	11-36 μmol L^{-1}	Hemochromatosis/anemia
	Urine	0.2-1.0 μmol L^{-1}	Desferrioxamine test
Lead (Pb)	Blood	<1.2 μmol L^{-1}	Lead poisoning
Lithium (Li)	Serum	0.6-1.5 mmol L^{-1}	Therapeutic drug in bipolar mood disorder
Magnesium (Mg)	Serum	0.7-1.0 mmol L^{-1}	Bone disease
Manganese (Mn)	Serum	35-118 nmol L^{-1}	Manganese deficiency
Mercury (Hg)	Urine	<0.5 μmol 24 h^{-1}	Mercury poisoning
Potassium (K)	Serum	3.2-4.8 mmol L^{-1}	Electrolyte disturbances
Sodium (Na)	Serum	132-148 mmol L^{-1}	Electrolyte disturbances
Zinc (Zn)	Serum	12-18 μmol L^{-1}	Acrodermatitis entero-pathica

of the element can be of value to the clinician. Colorimetric methods exist for many of the elements shown in Table 1, although AAS is the method of choice for most of them. Calcium, iron, and magnesium are routinely assayed in many laboratories by colorimetric methods, while sodium and potassium in serum are more commonly measured by flame emission spectrometry or increasingly by ion-selective electrodes.

One of the prime considerations to a clinical chemist in the selection of an appropriate method is sample size. The amount of available sample is often severely limited, and any method chosen should have as low a sample consumption as possible. A rapid throughput is also often required, and ideally a chosen method should be easy to automate. Most assays are carried out by diluting the specimen, manually or by semiautomatic diluters, and then

TABLE 2   Elements Determined in Body Fluids Using an FIA-AAS
System

Element	Body fluid	Volume (μL)	Throughput (assays h⁻¹)	Wavelength (nm)	Ref.
Calcium	Serum	4	120	422.7	8
	Serum	5		422.6	9
	CSF	5	120	422.3	10
Cobalt	Serum	100	120	240.7	11
Copper	Liver (bovine)	Digest	180	Not given	19
	Serum	120	120	324.8	12
	Whole blood	120	80	324.7	13
	Parotid saliva	50	240	324.8	14
	CSF	100	120	324	10
Iron	Serum	150	120	248.4	15
	Whole blood	100	80	217.0	16
	Parotid saliva	100	240	248.3	14
	CSF	100	120	248.3	10
Lead	Urine	Extract	—	217.0	16
Lithium	Serum	10	120	670.8	17
Magnesium	Serum	4	120	285.2	8
	Serum	20		202.6	9
	CSF	5	120	285.2	10
Manganese	Serum	100	120	279.5	11
Potassium	Serum	5	100	766.5	18
	CSF	5	120	789	10
Sodium	Serum	5	100	330.3	18
	CSF	5	120	589	10
Zinc	Liver (bovine)	Digest	180	213.9	19
	Serum	120	120	213.9	12
	Whole blood	60	80	213.9	13
	Parotid saliva	100	240	213.9	14
	Serum[a]	100	180	213.9	20
	Serum	100	40	213.4	21
	CSF	20	120	213.5	10

[a]Protein-free filtrate.

introducing it into the nebulizer in the case of flame atomic atomization or into an electrothermal atomizer.

Flow injection analysis (FIA) coupled with atomic absorption detection offers many advantages over previous techniques.   Since its
original conception by Růžička and Hansen (2) and Stewart et al.

(3) in the mid-1970s, FIA has been growing steadily in popularity. However, clinical chemists have so far been slow to appreciate the advantages of the technique. The application of FIA to clinical chemistry was reviewed by Rocks et al. (4,5). The earliest report of FIA with atomic absorption detection appears to have been by Zagatto et al. in 1979 (6). In 1983 Tyson (7) summarized the literature on the combination of flow injection analysis/atomic absorption spectrometry (FIA-AAS) and at that time there were only three papers on the use of FIA-AAS to measure elements in body fluids.

In recent years there has been an increase in the number of papers describing the use of FIA-AAS in a clinical setting. The elements that have been assayed in body fluids by FIA-AAS are shown in Table 2. It can be seen that FIA methods have been developed for nearly all the elements listed in Table 1, the exception being those elements present in ultratrace amounts that are normally measurable only by the use of electrothermal atomization and mercury for which cold vapor atomization is used.

2 FIA-AAS METHODS FOR CLINICALLY RELEVANT ELEMENTS

2.1 Copper and Zinc

The first use of FIA-AAS to measure elements of clinical interest in biological samples was the determination of zinc and copper in bovine liver by Wolf and Stewart in 1979 (19). They used the automated multiple flow injection analysis (AMFIA) system they had described previously (22) to assay both elements in NBS Standard Reference Materials, including SRM-1577, which is derived from bovine liver. Their flow system consisted of an automatic sampler, a sample insertion valve (constructed from four four-way slider valves), a sample withdrawal pump, and a depulsed positive-displacement solvent pump connected via small-bore Teflon tubing to the nebulizer of an atomic absorption spectrometer. A range of sample volumes (25 to 300 μL) were used for the various food digests and standard reference materials assayed. Good precision was obtained (2% coefficient of variation CV) at an acceptable throughput of 120 to 180 samples per hour.

Copper and zinc have been the elements most frequently determined by the combination of flow injection analysis and atomic absorption spectrometry, having been assayed in serum (12,20), whole blood (13), and parotid saliva (14). These trace elements are of particular importance in several inherited diseases (23). Acrodermatitis enteropathica is a severe zinc-deficiency state seen in infancy where the child has severe diarrhea, dermatitis, and alopecia. The condition is attributable to a defect in zinc absorption from the

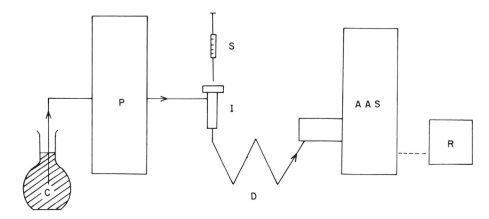

FIGURE 1 Manual injection flow injection analysis/atomic absorption
spectrometry. C, carrier solution (demineralized water); P, single-
piston HPLC pump; S, Hamilton microsyringe (100 μL); I, septum
injector port; D, dispersion tube (10 cm of 0.5-mm-ID PTFE); AAS
Rank Hilger Atomspek H1550 atomic absorption spectrometer; R,
chart recorder. Carrier is pumped at 3.9 mL min^{-1} in the direction
of the arrows. Serum (100 μL) is injected through a rubber septum
at I and is then swept through coil D into the nebulizer of the Atomspek. [Reproduced with permission of Blackwell Scientific Publishing (12).]

gut, resulting in zinc depletion with serum zinc concentration falling to 5 to 6 μmol L^{-1} (normal range 12 to 18 μmol L^{-1}). Clarifica-
tion of the role of zinc in the occurrence of acrodermatitis enteropathica is leading not only to advances in treatment but also to a
better understanding of the many roles of zinc in human metabolism.
Zinc is known to have an important involvement in the process of
tissue repair and wound healing, as it is necessary for DNA synthe-
sis. Because of this, serum zinc is often measured in patients with
chronic ulcers or certain gastrointestinal disorders that require surgery, for instance Crohn's disease (24). Many of these patients
are found to have low serum zinc concentrations because of prolonged malabsorption due to their intestinal dysfunction. These
patients often receive parenteral nutrition by intravenous line after
surgery and unless they receive zinc supplementation can rapidly
become zinc deficient. Zinc measurements are also commonly carried
out on other patients receiving long-term parenteral nutrition.

    The measurement of copper is of diagnostic importance in Wil-
son's disease, where there is a genetic deficiency of the copper-
binding protein, ceruloplasmin. Normally, absorbed copper is first
attached to albumin in the blood and then taken up by the liver.

Subsequently it is released, bound to ceruloplasmin. In the absence of its carrier protein copper is carried loosely bound to albumin, which allows excessive copper to be deposited in tissues and increases urinary copper excretion. The high tissue levels predominantly affect the liver, resulting in cirrhosis, and the brain, causing progressive involuntary movements and disability. The disease can be diagnosed by demonstrating reduced ceruloplasmin, low serum copper, and increased urinary copper (25). Menke's "kinky hair" syndrome is associated with an apparent copper deficiency (26), although at present the mechanism for the disease is poorly understood.

Rocks et al. (12) determined zinc and copper in serum using FIA-AAS with a manual injection system. The flow system is shown in Fig. 1. Serum (100 µL) was injected, by means of a Hamilton microsyringe through a rubber septum, into a flowing stream of deionized water supplied by a single-piston high-performance liquid chromatographic (HPLC) pump. The atomic absorption instrument used was a Rank Hilger Atomspek H1550. Using a flow rate of 3.9 mL min^{-1}, a sample throughput of 120 samples per hour was obtained with a precision of 1.7% CV for zinc and 2.1% CV for copper. The FIA-AAS methods were compared to conventional manual methods, which required a 1:1 dilution of serum with trichloroacetic acid (TCA) to precipitate the serum proteins. The supernatant could then be aspirated into the nebulizer of the Atomspek until a steady-state signal was achieved. Good correlations were obtained between the two methods (r = 0.97 for zinc and r = 0.98 for copper).

Although these flow injection methods functioned adequately and reduced sample requirement from 1 mL to 100 µL for each element, they retained some of the disadvantages of manual methods. Subsequently, a rotary valve was used to introduce the sample, as shown in Fig. 2. The rotary valve was fitted with a variable sample loop, allowing the sample volume to be adjusted up to 100 µL. The serum sample was drawn through the valve, which had been turned to the fill position, using a peristaltic pump. Sufficient sample was drawn in to fill the loop completely and provide an overlap on either side of the valve; this resulted in a sample requirement of approximately 150 µL. The peristaltic pump was then stopped and the valve turned to introduce the measured "slug" of sample contained within the loop into a stream of demineralized water. The sample was then swept into the nebulizer via a short (10 cm) dispersion tube [0.5-mm-ID polytetrafluoroethylene (PTFE)]. These conditions resulted in a dispersion D of 1.2. To eliminate the need for the pump supplying the carrier stream to be stopped when the valve was in the "fill" position, a bypass loop was provided around the valve, enabling a continuous flow of carrier and producing a more stable baseline than if the flow was stopped and started

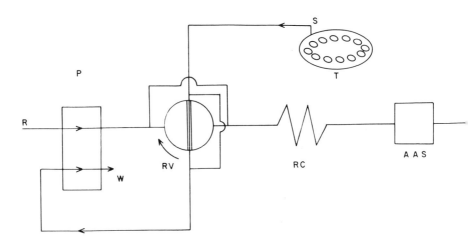

FIGURE 2   Incorporation of rotary valve into FIA-AAS system.
R, reagent (demineralized water); P, 8 roller peristaltic pump; W,
waste container; RV, rotary valve with variable sample loop; RC,
reaction coil (10 cm of 0.5-mm-ID PTFE); AAS, Rank Hilger Atomspek H1550 atomic absorption spectrometer; S, samples; T, autosampler turntable.   Reagent is constantly pumped through the system to the nebulizer of the Atomspek.   When the rotary valve is
turned to the "fill" position, as shown, the reagent flow passes
round the bypass loop to ensure a continuous supply to the nebulizer.   With the valve in this position the sampler probe dips into
the sample and serum is drawn through the valve until the valve
chamber is full.   The valve is then turned through 90°, introducing a slug of sample into the reagent stream.   The sample probe
returns to a wash chamber and demineralized water is flushed
through the valve loop to remove the residue of the previous sample.   [Reproduced with permission of Taylor & Francis Ltd. (67).]

whenever a sample was introduced into the system.   The peaks obtained from a run of standards and serum samples are shown in
Fig. 3.

Standardization was performed using standards prepared in a
viscosity-adjusted solution containing sodium, potassium, chloride,
and albumin.   The need to use viscosity-adjusted standards when
measuring elements in body fluids by FIA-AAS is particularly pronounced when the dispersion is reduced to values of D of less
than 10.

Attiyat and Christian have also measured zinc in serum by FIA-
AAS (20).   In their analytical system no pump was used, the deionized water carrier stream being introduced into the nebulizer of their
atomic absorption instrument by utilizing the negative pressure generated by the nebulizer.   They pretreated their serum samples either

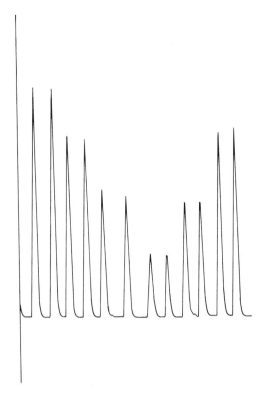

FIGURE 3 Peaks obtained from FIA-AAS method for zinc in serum. The FIA-AAS system as shown in Fig. 2 was used with a sample volume of 100 μL. Peaks from left to right in duplicate are: serum 1, serum 2, 5, 10, 15, and 20 μmol L^{-1} viscosity-adjusted zinc standards.

by dilution 1:5 with demineralized water or by protein precipitation using a 1:1 mixture of serum and 25% TCA (w/v), followed by centrifugation for 5 min. They compared the results obtained by these two methods and found that direct dilution produced results slightly higher than those using a protein-free filtrate. The FIA-AAS method using prior protein precipitation agreed well with direct aspiration of the same protein-free filtrates, and a comparison of the standard curves showed that the FIA-AAS system had a dispersion D of approximately 1.1. The finding of higher results with specimens that had not had the serum proteins removed could be related to viscosity effects, as the standards used were not viscocity adjusted.

Burguera et al. measured zinc and copper in various body fluids using an FIA-AAS system based on that described by Fukamachi

and Ishibashi (27), who had used it with organic solvents. Burguera et al. measured zinc, copper, and iron in parotid saliva by
injecting varying volumes of saliva (50 μL for copper, 20 μL for
zinc, and 100 μL for iron) via a rotary valve without pretreatment
of the samples (14). Analytical precision was adequate with within-
batch variations of between 1.8 and 2.2% (CV) and between batch
variations of 2.9 to 4.4%. Recoveries for all three elements ranged
from 97 to 101%.

Using the same system Burguera et al. have measured zinc
and copper in microvolumes of CSF (10) and in human vitreous hu-
mor (the clear viscose fluid obtained from the eye) (28). Measurement of the metal content of whole blood by atomic absorption spectrometry usually involves prior treatment of the sample to lyse the
erythrocytes and white cells of the blood, followed by assay of the
resulting solution for the required elements. If the analysis is to
be carried out by flame atomic absorption spectrometry, it is important to ensure that the content of particulate matter in a specimen
is kept as low as possible to avoid nebulizer capillary tube blockage.
This is usually achieved by centrifugation of a mixture of sample
and lysing agent to separate the red cell "ghosts" and other cell
fragments.

Burguera et al. (13) have applied the technique of microwave
sample dissolution in an FIA system to the preparation of whole
blood for the measurement of iron, copper, and zinc. Using a double injection valve, sample (90, 60, and 100 μL of whole blood for
copper, zinc, and iron, respectively) was introduced into a stream
of carrier solution and acid reagent being introduced into a second
stream. The two streams were merged a short distance downstream,
after which the reaction tube was connected to 50 cm of coiled Pyrex
tube which was contained within a domestic microwave oven. While
the sample/reagent mixture was flowing through the dissolution tube,
the microwave heating acted as a source of energy, rapidly heating
the sample, which in combination with the acid reagent promoted
mineralization of the blood. Detergent (Triton X-100) was added to
the carrier solutions to avoid clogging the tubing leading to the
oven. Using a flow rate of 1.8 mL min^{-1}, the sample was resident
within the oven for between 15 and 25 s; below 15 s mineralization
was incomplete, while above 25 s the carrier solution flow pattern
was affected by the evolution of gas from acid fumes generated during dissolution. The resulting solution could be introduced directly
into the AAS nebulizer without clogging of the nebulizer or burner.
Analytical performance was satisfactory and the method provides a
simple means of determining relevant metallic elements in whole blood
without prior ashing, digestion, or precipitation of samples at a
throughput of 80 samples per hour.

A group from Denmark (21) have produced an automated FIA-AAS method for zinc in human serum. Their system does not contain a pump, but relies on the negative pressure of the nebulizer to draw solution through the instrument in a manner similar to that of Attiyat and Christian (20). Samples were placed in an autosampler and aliquots drawn through a single three-way valve which was under microcomputer control. The serum samples were measured directly without pretreatment and blanks were run between samples to avoid carryover. Two aliquots of sample of 50 µL each were used, the first being discarded as it was slightly diluted by contact with the carrier solution left in the sample tube from the blank. The sample volume was controlled by the length of time the valve was open when the probe was in the sample, which could be varied via the computer software. Both peak height and peak area measurements were taken, the latter allowing calibration with aqueous standards, while the former required matrix-matched standards due to viscosity effects. Accuracy and precision were acceptable for routine use (between-batch CV was approximately 5%). The system is a fairly simple means of automating an FIA-AAS method but suffers the disadvantage of wasting 50% of the sample volume.

2.2 Iron

Iron, one of the earliest elements to be recognized as being essential to humans, has been determined by FIA-AAS in serum (15), saliva (14), CSF (10), and whole blood (13). Iron absorbed from the diet is transported in plasma bound to the specific carrier protein transferrin. Normally only about one-third of transferrin is bound with iron, representing 30% saturation of total iron binding capacity. The majority of the body's iron stores are located in the liver. Iron deficiency may result from inadequate intake, poor absorption, or excessive blood loss. Hemochromatosis is a condition caused by excessive accumulation of iron in tissues and organs. The mechanism of the increased total body iron concentration is a genetically determined excessive absorption of iron from the gut. The organs principally affected include the liver and pancreas, which can be badly damaged by the progressive deposition of iron. Treatment is by removal of the iron using a chelating agent such as desferrioxamine or by regular venesection (bleeding). The effect of chelation therapy can be monitored by measuring the return toward normal of the increased urinary iron excretion (29).

Rocks et al. (15) determined serum iron and total iron-binding capacity using a modification of the system shown in Fig. 2. Serum iron could be measured using a rotary valve with a sample loop of 150 µL. A good correlation (r = 0.987) was obtained for samples assayed by both the FIA-AAS method and a colorimetric method on a Technicon AA1 AutoAnalyzer. The FIA-AAS method had a within batch precision of 1 to 2% (CV) and a between-batch CV of 3 to 7%. Colorimetric methods for iron are not sensitive, and to achieve adequate absorbances relatively large volumes of sample are required; the Technicon AutoAnalyzer method consumed at least 1 mL of sample.

Some indication of the available iron-binding capacity of serum is of importance in the assessment of anaemia, and the most commonly measured parameter is the total serum iron binding capacity (TIBC), which acts as an indirect assessment of transferrin concentration. The measurement of TIBC using FIA-AAS was achieved by prior treatment of 200 μL of serum with 400 μL of acidic ferric chloride solution to saturate all the transferrin-binding sites. Any excess iron was then removed by the addition of magnesium carbonate (40 mg), which precipitated any unbound iron. After centrifugation, the supernatant was treated as a normal serum and after protein precipitation was injected into the FIA-AAS system. From the serum iron concentration and the TIBC result, the percent saturation of the serum could be calculated; in healthy people the range of percent saturation normally encountered is 30 to 45%, while in anemia the value can fall as low as 6%.

2.3 Lithium

Lithium is not considered to be an important serum cation, as it is only present in low concentrations. It is, however, used increasingly in the management of manic-depressive patients (the bipolar mood disorders), to whom it is given in the form of lithium carbonate. Doses of as much as 1.2 g per day are often used, and there is a narrow gap between therapeutic plasma concentrations and levels sufficient to cause toxicity. To avoid side effects and yet maintain a plasma concentration within the therapeutic range (0.6 to 1.5 mmol L^{-1}, 12 h after a dose) it is important to monitor the serum concentration of patients receiving this drug on a regular basis.

Flame emission photometry (31) and atomic absorption spectrometry (32) are the most commonly used techniques for lithium estimation, and there is no definite advantage of either method. Flame emission is generally more sensitive than atomic absorption spectrometry for lithium determination, but is also less precise (33). Serum lithium has been determined using the simple system shown in Fig. 1 (17). Demineralized water was used as the carrier solution and serum (10 μL) was introduced by microsyringe through a rubber septum into the flowing stream. Using a flow rate of 4 mL min^{-1} and 25 cm of 0.5-mm-ID tubing as a dispersion tube, a D value 12 was obtained. This was sufficient to give an absorbance reading of 0.25 A for a 1 mmol L^{-1} lithium standard, allowing the construction of a linear standard curve from 0 to 2.0 mmol L^{-1}.

It was found that the standards must contain physiological concentrations of sodium, potassium, and chloride, as addition of 140
mmol L^{-1} of sodium chloride enhanced the signal obtained by 12%.
Clinically encountered variations in the concentrations of these ions
from their normal values produced little difference. Using suitably
matched standards, a correlation of r = 0.996 was obtained between
the FIA method and a conventional atomic absorption method. At a
throughput of 180 samples per hour a precision of 3% (between-
batch CV) was achieved. The injection system has subsequently
been replaced with a miniature rotary valve.

2.4 Calcium and Magnesium

These two elements have been measured in serum for many years
and have an established role in the diagnosis and treatment of many
disorders. Calcium is required for the control of many biochemical
pathways at a cellular level and acts as a coupling agent between
muscle excitation and contraction. Its whole body and serum concentrations are regulated by several different processes; absorption
through the gut, parathyroid hormone secretion, and vitamin D concentration. In plasma, calcium exists in the ionized state and is
also bound to albumin (34). Magnesium is also found in plasma in
ionized and protein-bound forms (albumin binding approximately
one-third of the magnesium in the plasma) (35). Magnesium is required as a cofactor by many of the enzymes involved in phosphate
transfer reactions utilizing adenosine triphosphate (ATP). Magnesium
measurement is important in the patient on hemodialysis, in the post-
operative patient, in patients receiving total parenteral nutrition
(TPN), in women with premenstrual syndrome (PMS), and in cases
of chronic gastrointestinal disturbance, where magnesium deficiency
can occur.

Although colorimetric methods exist for the measurement of both
calcium and magnesium in plasma, AAS is the reference method for
both elements and is used routinely in many laboratories, especially
for the assay of magnesium. Both of the elements absorb strongly
in the flame, so to obtain practicable sensitivity, serum must be
diluted at least 50-fold. For magnesium the dilution factor often
has to be as much as 100-fold, although this will obviously vary
with the particular instrument used.

Since the first measurements of calcium in serum by AAS, interference in the assay by protein, phosphate, sulfate, sodium, and
potassium have been reported (36-42). A variety of solutions have
been described to eliminate these interferences and have been tabu-
lated by Pybus et al. (42). A reference method for the determination of calcium in serum based on a 50-fold dilution of serum in hydrochloric acid (50 mmol L^{-1}) containing lanthanum chloride (10 mmol

L^{-1}) has been shown to be unaffected by known sources of interference (43). Magnesium is much less affected by interferences than calcium, and ethylenediaminetetraacetic acid (EDTA)-containing solutions have been proposed as suitable diluents (41).

The measurement by FIA-AAS of calcium and magnesium in serum using these diluents has been described. The method employed a miniature four-port rotary valve (Hamilton HVP, No. 86782) to inject a 4 µL of serum into a stream of appropriate diluent flowing at 3.5 mL min^{-1}. A 2-m coil of 0.8-mm-ID dispersion tube was found to be necessary to obtain the desired 50-fold dilution of the serum for the assay of calcium. This system produced a dispersion of about 54, which was, however, unsuitable for magnesium measurement, producing absorbances greater than 2.0. Lengthening the tube to 5 m still did not achieve a sufficiently high dispersion. However, increasing the internal diameter of the tube to 1.0 mm allowed a tube length of 2 m to be used, with a resulting dispersion of approximately 100. The methods using the FIA-AAS technique correlated well with alternative methods (o-cresolphthalein complexone colorimetric method for calcium and manual AAS for magnesium) and between-batch CV values of 0.87 and 2.27% were obtained for calcium and magnesium, respectively.

Burguera et al. (9) described an FIA-AAS method using manual injection of serum by a Hamilton microsyringe through a rubber septum. Their system did not use a pump, the carrier flow rate being controlled by adjustment of the airflow regulating valve on the nebulizer of their atomic absorption instrument. They used a short (20 cm) length of dispersion tube of unspecified internal diameter. Rocks et al. commented at the time (44) that this would produce only a low dispersion of sample, accounting for the nonlinear results obtained using the 285.2-nm absorption line of magnesium. Burguera et al. overcame this problem by changing to the much less sensitive absorption line at 202.6 nm and using a larger sample volume (20 µL rather than the 5 µL used for the assay of calcium). Creating a FIA-AAS system without a pump that can achieve a dispersion greater than about 50 would be difficult, as the negative pressure generated by the nebulizer would be insufficient to draw the carrier solution through the system at a reasonable flow rate. The variation of approximately 3.5% obtained by Burguera et al. for their pumpless magnesium assay is significantly greater than that obtained by Rocks et al. (8) for a pumped system. The CV of approximately 1.0% obtained for calcium determination in the pumpless system is, however, directly comparable to that found in a pumped system (8). In our experience, results are more reproducible if a pumped carrier stream is used than if the carrier is aspirated through the injection port, particularly if a long dispersion tube is employed.

2.5 Sodium and Potassium

Sodium is the most abundant extracellular cation, while potassium
is a major intracellular cation (45). Sodium is involved in the majority of the metabolic processes of the body, maintaining osmotic
pressure, hydrating body tissues, and aiding in the regulation of
the pH of the blood. The plasma sodium concentration changes in
many disease states. Potassium is vital in the maintenance of the
electrical charge balance across most cell membranes, and hypo- or
hyperkalemia can produce life-threatening cardiac arrhythmias. The
body concentration of both these ions is regulated primarily by the
kidney, and diminished renal function can lead to considerable ab-
normalities in the plasma concentrations of both. The large number and variety of pathological conditions that affect electrolyte
metabolism put sodium and potassium among the most commonly performed assays in the clinical laboratory.

   Burguera et al. have measured sodium and potassium in serum
using atomic absorption spectrometry by their pumpless flow injection system (18). Serum (5 μL for each electrolyte) was introduced
by the use of a Hamilton microsyringe. The absorption of sodium
was assessed at 330.3 nm and potassium at 766.5 nm. The system
functioned at a throughput of 100 measurements per hour with good
precision (CV between-batch values of 1.23% and 3.26% for sodium
and potassium, respectively).

2.6 Lead

Many of the heavy metals are toxic to humans and the detection and
quantitation of some of these elements can be important in both toxicology and environmental medicine. The most commonly measured
heavy metal is lead, which has a widespread industrial use, being
present in some paints, solders, storage batteries, print type, cer-
amics, and motor engine fuel. Blood lead measurement is often carried out on many workers involved in these industries, as high
levels can cause neurological damage and in some cases can prove
to be fatal (46). As 95% of the lead in blood is contained in the
erythrocytes, serum lead measurements are unsatisfactory for either
detection or quantitation of lead exposure. Whole blood lead determinations are usually performed by AAS with electrothermal atomization (47) or by flame AAS with the Delves cup (48). Urine lead
levels are sometimes used as a screening test, but the relationship
with whole blood concentrations is not sufficient to allow urine measurements to replace those in whole blood as a confirmation of the
diagnosis of lead poisoning (49).

   Burguera et al. have used FIA-AAS to determine lead in 24-h
urine specimens collected from workers on a printing press and from
nonoccupationally exposed individuals. They use their standard

FIA-AAS system fitted with a 20-μL Rheodyne rotary valve. The urine samples or standards were first treated with methyl isobutyl ketone (MIBK) with the addition of ammonium pyrrolidinedithiocarbamate (APDC) to extract the lead into the solvent, which is then injected into the flow system. The detection limit of the system was about 10 μg L^{-1}, with a linear range of 15 to 200 μg L^{-1}. Using this technique, they found a significantly higher urinary lead excretion (72 ± 26 μg L^{-1}) in the print workers than in the unexposed group (39 ± 8 μg L^{-1}).

2.7 Cobalt and Manganese

In recent years there has been an expansion of interest in the metals present in the body in trace or ultratrace amounts. Cobalt is known to be required by humans in microgram amounts for the formation of vitamin B_{12}, which is a cyanocobalamin complex (50); however, it is also toxic if sufficient amounts are taken. Manganese is essential for bone and tissue formation, carbohydrate metabolism, reproductive processes, and lipid metabolism (51). Both metals are typically determined using furnace AAS or neutron activation analysis. However, Burguera et al. (11) have reported the use of their flame FIA-AAS apparatus to determine both elements in serum directly. Using microsyringe injection they obtained results from the injection of serum (100 μL) into a flowing stream of water. Standards were viscosity adjusted.

3 DISADVANTAGES OF CONVENTIONAL FIA-AAS

The variety of elements that have been measured by FIA-AAS in various body fluids is a testimony to the benefits of flow injection analysis over existing sample introduction techniques for AAS. Several disadvantages remain with the various methodological approaches described so far in this chapter. Manual injection with a microsyringe through rubber septa is difficult to mechanize without expensive HPLC autosampler equipment. Nebulizer blockages occur with greater frequency when septa are used, due to pieces of rubber being cut out from the septa by the repeated passage of the needle. Both these problems can be eliminated by replacement of the injection septum with a rotary valve that can be filled either manually with a syringe or by the use of a separate pump for the sample line or by an automatic system utilizing an autosampler (as shown in Fig. 2). The latter alternative allows easier automation and removes the risk of contamination of the sample from the metallic needle of most microsyringes (this is particularly troublesome when assaying the ultratrace elements, e.g., cobalt, chromium, and manganese).

Rotary or slider valves, however, introduce several disadvantages of their own. To ensure accurate metering of the sample volume by the sample loop of the valve, extra sample must be drawn through the valve to fill the sample lines on either side of the valve. Unless a wash solution is put through the valve between samples, the extra sample must also flush out the residue of the previous sample. This requirement can result in a sample consumption of at least twice the volume actually delivered to the nebulizer of the atomic absorption instrument. Valves are also prone to leak or stick when exposed to the heavy work load of a routine clinical chemistry laboratory. It was to remove these problems that Riley et al. developed a variant of FIA [which they called controlled dispersion analysis (CDA)] that dispenses with the requirement for injection septa or valves and yet consumes minimum quantities of sample and reagent (52). The technique has subsequently been coupled in atomic absorption spectrometry to give a sample introduction system with great flexibility for use in clinical chemistry laboratories (53).

4   CONTROLLED DISPERSION ANALYSIS

In conventional continuous flow analysis the sample and reagents are aspirated by the action of a peristaltic pump into a flowing stream of liquid interspersed with air bubbles. The pump runs continuously drawing reagent through polyvinyl chloride tubes with variable internal diameters to permit the selection of differing amounts of reagents. When a sample is offered to the system, a probe moves from a wash chamber into the sample cup. The probe remains in the sample for a fixed period of time (set via a cam timer), resulting in a measured amount of sample being drawn up into the sample pump tube. This method of sample introduction is not applicable to FIA, as while the probe is in the air the pump is still running, causing a small air bubble to be drawn into the system. However, if the peristaltic pump could be stopped and started easily and controlled so as to make precise and reproducible angular movements, a precise volume of sample could be aspirated without introducing air into the system. Provided that the slug of sample produced in this manner would pass through the pump without excessive or variable dispersion, the system should have the benefits of FIA without the disadvantages of injection valves or septa.

4.1   Instrumentation

The prime requirement of a flow analysis instrument based on the foregoing principle is a pump which can be started and stopped quickly with fine control of its rotation. The pump used had eight closely spaced rollers, 9 mm in diameter, moving over a curved

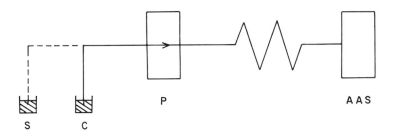

S   C

FIGURE 4   Flow system for controlled dispersion analysis. S, sample cup; C, carrier solution (demineralized water); P, eight-roller peristaltic pump; AAS, Rank Hilger Atomspek H1550 atomic absorption spectrometer.   The probe normally resides in carrier solution, C; when a sample is present, the pump is stopped and the probe is transferred to the sample container.   The pump then rotates to aspirate sample, stops to allow the probe to return to the carrier solution, and then restarts, pumping the sample "slug" through the dispersion coil to the nebulizer of the atomic absorption instrument.   [Reproduced with permission of the Royal Society of Chemistry (53).]

platen (development model supplied by Corning Medical and Scientific, Sudbury, Suffolk, United Kingdom).   These had spaces for up to eight pump tubes.   To give the control of movement necessary, the pump was driven by a stepper motor (Impex ID 29, which has a 3.5° step angle and a pull-in torque of 0.25 N m; Dennard Rotadrive, Fleet, Hants, United Kingdom).   The pump was coupled indirectly to the stepper motor by a pair of gears, giving a 2:1 reduction.   Control of the stepper motor was provided by a Cromemco Z2D microcomputer with two 5.25-in. floppy disk drives.   The control program was written in a mixture of BASIC (parameter selection and data handling) and machine code (motor control).

The basic flow system is shown in Fig. 4.   A probe connected to a pump tube resided in a reagent or carrier solution container, permitting carrier to be pumped slowly through the system when there were no samples for analysis.   At the start of each sample cycle the pump was indexed to its start location (it was found that precision could be improved by starting each cycle from the same pump roller); the pump was then halted and the probe was transferred from the carrier solution to the sample container.   The pump was then restarted, rotating through a small angle, drawing sample into the probe.   The pump was halted while the probe was returned to the carrier solution.   With the probe back in carrier solution/reagent, the pump was restarted, propelling the slug of sample through the pump and onward along the reactor tubing to the detector.

The probe-transfer mechanism was a simple cam-operated device that raised or lowered an arm and rotated it through 90°. Sample [and reagent(s) where appropriate] probes made of 0.3-mm-ID stainless steel were attached to this arm. The length of the reactor tubing and its internal diameter was used to vary the dispersion and to ensure adequate mixing of sample with reagent. Polytetrafluoroethylene (PTFE) tubing (Omnifit Ltd., Cambridge, United Kingdom) of various internal diameters (0.4 to 1.0 mm) and lengths (30 cm to 2 m) was used for the majority of applications. However, the simplest way to vary the dispersion is by altering the aspirated sample volume. The volume of sample could be controlled with considerable precision by varying the number of steps taken by the motor while the probe was in the sample. As this can easily be programmed from the microcomputer, volumes from 240 nL upward have been used.

4.2 CDA-AAS

Initially, the controlled dispersion analyzer was arranged as shown in Fig. 4. A single pump tube was used (0.74-mm-ID) connected at one end to 30 cm of PTFE tubing (0.4-mm ID). This was attached to the nebulizer of a Rank Hilger H1550 Atomspek (Rank Hilger, Margate, Kent, United Kingdom). The nebulizer aspiration rate was determined to be 3.5 mL min^{-1} and the pump was adjusted to match this flow rate. This simple arrangement, which had worked well with colorimetry, proved to be unsatisfactory when applied to some atomic absorption analyses. The baseline signal was erratic and the precision obtained in early experiments measuring lithium in serum was poorer than that achieved using conventional FIA-AAS with injections performed via a rotary valve. Further experiments identified the problem as being due to the requirement of halting the pump while the probe was in air moving between the sample and reagent containers. During this period the nebulizer of the atomic absorption spectrometer was allowed to run dry. A similar problem has been described by Yoza et al. (54) in an assay for magnesium using conventional FIA with a rotary valve. Their solution was to incorporate a T-piece after the pump, to which is connected a piece of tubing leading to a separate reservoir of carrier solution. With this adaptation, when the pump is stationary, the nebulizer will aspirate from this container. A T-piece was incorporated into the CDA-AAS system as shown in Fig. 5. The flow rate produced by the pump required adjusting to ensure that it did not exceed the nebulizer aspiration rate. If the pump flow rate was faster than the nebulizer aspiration rate, flow occurred not only along the main channel but also along the tubing into the wash reservoir. If a sample had been introduced, not only was a variable amount lost along this secondary

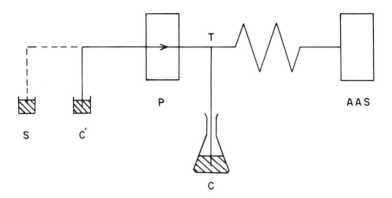

FIGURE 5 Incorporation of carrier reservoir into the CDA-AAS
system. S, sample cup; C', reagent reservoir; C, wash reservoir;
P, eight-roller peristaltic pump; T, T-piece (Elkay Labs LP2); AAS,
Rank Hilger Atomspek H1550 atomic absorption spectrometer. The
CDA system is as described in Fig. 4. When the pump is stopped
while the probe is in the air moving between the sample and reagent containers, S and C', the negative pressure of the nebulizer
draws wash solution from container C, ensuring that the nebulizer
does not run dry. [Reproduced with permission of the Royal Society of Chemistry (53).]

channel, but the wash reservoir was itself contaminated. A slower
pump flow rate resulted in the nebulizer drawing some solution from
the wash reservoir, the only effect being to increase the dispersion
of the system. Thus, provided that the pump flow rate was less
than or equal to that of the nebulizer, the atomic absorption flame
did not "run dry" and a more stable baseline signal was obtained.

     For the assay of calcium and magnesium, this modification pro-
vided an added benefit. Both assays require the use of a reagent
such as EDTA or lanthanum chloride. Pumping these solutions continuously for any length of time results in increased burner clogging with a subsequent reduction in reproducibility. If this reagent was used only in the carrier reservoir and water was placed
in the wash reservoir, the reagent was pumped only when a sample
was in the instrument, with water being drawn up from the wash
reservoir at all other times. A wash stage could therefore be incorporated into each sample cycle by stopping the pump after the peak
had passed through the flame and the result obtained. The nebulizer then draws water through the dispersion coil, the nebulizer,
and the burner, flushing any residual sample and reagent out. By
varying the delay time before another cycle was initiated, an appropriate wash time could be selected for any particular assay.

TABLE 3 Analytical Characteristics of the CDA-AAS Methods

Element	Sample size (μL)	Reagent/water consumption (mL)	Cycle time (s)	Linearity
Lithium	13.5	—/1.7	20	To 2 mmol L^{-1}
Magnesium	2.5	0.5/1.2	25	To 1.5 mmol L^{-1}
Zinc	120	—/1.6	30	To 40 μmol L^{-1}
Copper	120	—/1.6	30	To 40 μmol L^{-1}
Calcium	2.5	0.5/1.2	25	To 3.5 mmol L^{-1}

The methods previously described for serum lithium (17), zinc (12), copper (12), calcium (8), and magnesium (8) by FIA-AAS were then reexamined using CDA-AAS. The features of the CDA-AAS methods are shown in Table 3. The methods were compared to the FIA-AAS methods and the within-batch precision determined for each analyte. The comparison data are shown in Table 4.

The most significant advantage of CDA-AAS is the ease with which the sample volume can be varied. The dispersion coefficient, D, may be varied either by changing the volume injected or by using dispersion tubes of different lengths, internal diameters, or both. The flow rate cannot be altered over a wide range, as in FIA with other forms of detection, without a reduction in nebulizer performance. In a conventional FIA system, changes in D can be achieved only by physically altering the system in one of these ways. In CDA-AAS the sample volume can easily be varied over a very wide range without resorting to changing sample loops. The same tube manifold can therefore be used to produce an almost infinite number of dispersion values. In the methods shown in Table 3, the calcium

TABLE 4 Comparison of the Precision and Accuracy of the CDA-AAS Methods with FIA-AAS (n = 10)

Element	Within-batch precision (% CV)		Correlation	
	CDA-AAS	FIA-AAS	Slope	r
Lithium	1.8	2.9	1.04	0.96
Magnesium	1.18	2.27	1.01	0.98
Zinc	0.92	1.7	0.97	0.98
Copper	0.96	2.1	0.98	0.97

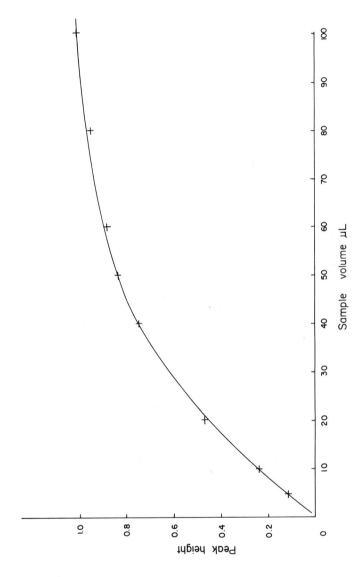

FIGURE 6   Relationship of sample volume with absorbance. Sample volumes of 5 to 100 μL of serum were sampled using the CDA-AAS method for the determination of magnesium.

and magnesium assays required a D value of about 80, while for the lithium assay D was 15 and for copper and zinc, 1.7. All these were achieved with the single manifold described earlier.

The ability to vary the sample volume allows standard calibration in cases where a suitable range of sample volumes can be found to produce a calibration graph over the desired range for that analyte. For example, the serum magnesium assay was usually calibrated by the introduction of 2.5 μL of 0.5, 1.0, 1.5, and 2.0 mmol L^{-1} standards. The same results were achieved by using 1.25, 2.5, 3.75, and 5.0 μL of a 1 mmol L^{-1} standard. As CDA has been used to measure albumin colorimetrically on only 240 nL of serum with a 2 to 3% CV (55), the only limitation to the use of the standard calibration would appear to be when the dispersion of the system is low and a further increase in injected volume results in a nonlinear response, as shown in Fig. 6.

The technique of "merging zones," first described by Bergamin et al. (56) and used by them for the determination of calcium and magnesium in plants by FIA-AAS (6), allows a considerable reduction in reagent volume. This technique can be applied equally well to CDA with the addition of a reagent probe(s) and a T-piece (Elkay Laboratories, Basingstoke, Hants, United Kingdom), as shown in Fig. 7. Synchronization of the sample and reagent slugs can be achieved by adjusting the relative lengths of the pump tubes prior to the T-piece until both sample and reagent arrive at the detector simultaneously. An inexpensive reagent, such as deionized water, is used as the carrier solution for both streams. This technique is most valuable when one or more of the reagents used in the reaction is expensive, as is the case for reagents containing enzymes for the specific colorimetric determination of analytes such as uric acid or cholesterol. The majority of reagents used in the FIA-AAS applications described in this chapter are relatively inexpensive and have not warranted the additional complexity of merging zones.

Merging zones could, however, be useful in a system similar to that described by Nord and Karlberg (57) for preconcentration of samples prior to their introduction into the nebulizer of an atomic absorption instrument. These workers produced a manifold which allowed the extraction of metal ions in aqueous samples (copper, nickel, lead, and zinc) into 4-methyl-2-pentanone with ammonium pyrrolidinethiocarbamate as an extracting agent. A membrane phase separator was employed to remove a slug of organic phase containing the extracted metals and introduce it into a water carrier line, in which it is not dispersed, giving a preconcentration of sample.

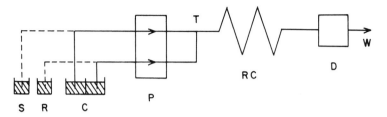

S   R   C

FIGURE 7 Merging-zone CDA. S, sample cup; R, reagent container; C, carrier solution; P, eight-roller peristaltic pump; T,
T-piece; RC, reaction coil; D, detector; W, waste. In a merging
zone CDA setup both sample and reagent probes are normally in a
carrier solution until a sample is presented to the instrument. The
pump is then stopped and both probes transferred to their respective sample and reagent vessels. The pump is then rotated through
a small angle, drawing small volumes of both sample and reagent into the pump tubes. The relative size of these volumes was adjusted
by using pump tubes with different internal diameters. The pump
is stopped, the probes returned to the carrier reservoir, and the
pump restarted. The "slugs" of sample and reagent travel along
their respective tubes until meeting at the T-piece. Synchronization is achieved by cutting the lengths of the two tubes until the
slugs meet. The diagram shows the standard CDA system for use
with a colorimetric detector; for AAS the modifications shown in
Fig. 5 are incorporated. [Reproduced with permission of the American Association for Clinical Chemistry (52).]

5 SPECIALIZED APPLICATIONS

All the applications discussed so far have utilized the advantages
of FIA as a means of pretreating samples before their introduction
into the nebulizer of an atomic absorption instrument. All have,
therefore, used a flame (air-acetylene or nitrous oxide-acetylene)
to achieve atomization. Several of the elements of clinical interest
listed in Table 1 (cadmium, aluminum, and mercury) cannot be assayed in clinical samples using flame atomization. Although the
combination of FIA with flame AAS has attracted the attention of
clinical chemists, the use of FIA with hydride generation, cold vapor atomization, or a graphite furnace has been restricted to analytical chemists.

5.1 Cold Vapor Atomic Absorption: Mercury

Mercury is one of the heavy metals toxic to humans in relatively
low concentration. Mercury poisoning leads to a variety of symp-
toms, including those related to gastrointestinal disturbances,
diarrhea, and dehydration. Monitoring of persons exposed to

mercury or mercuric compounds is most easily performed by measuring the mercury concentration in a sample of urine. The measurement of urine mercury is usually done by cold vapor atomic absorption, in which borohydride ions cause the chemical reduction of mercury ions to elemental mercury, which is then swept by an inert carrier gas into an absorption cell in an atomic absorption instrument, where its absorption at 253.7 nm can be measured. Most methods require manual introduction of both the sample and borohydride reagent into the reaction vessel, and relatively few automated procedures have been described.

A method for the determination of mercury in acidic standard solutions has been described which uses FIA in combination with a Teflon membrane to form a cold vapor atomic absorption system (58). The method uses the permeability of commercial Teflon [polytetrafluoroethylene (PTFE)] tape to elemental mercury, to form a gas-liquid phase separation system. A sodium borohydride solution and an aliquot of a mercury standard are introduced into the FIA manifold, which uses "merging zones" by means of a homemade proportional injector. The resulting Hg/H_2 generated by the reduction process permeates through the Teflon membrane, where its absorption is immediately measured. The major advantage this offers is the elimination of the reaction vessel and the automation of the process. This technique has not yet been applied to clinical samples but should be of interest to any laboratory with a high demand for urine mercury assays.

5.2 Hydride Generation: Bismuth

Bismuth compounds have been used to treat various gastrointestinal conditions, particularly in patients with a colostomy or ileostomy. Severe neurological disturbances sometimes occur in patients who have received prolonged oral administration of bismuth compounds. Blood bismuth levels would appear to be important in assessing these cases. Bismuth has always been difficult to measure, and hydride generation is probably the method of choice. Few automated hydride generator systems have been described in the literature.

Åstrom published a method in 1982 for the determination of bismuth in standard solutions by FIA-AAS with hydride generation (59). A sample, usually 700 μL was injected via a pneumatic valve into a line containing hydrochloric acid, and after mixing with sodium borohydride, the solution was sprayed with nitrogen or argon into the gas-liquid separator. The gaseous hydride was then swept into an electrically heated tube furnace and its absorption measured at 223.1 nm. A major advantage of the method was the reduction in the effect of interferents because of the improved control over the reaction time, which could be kept sufficiently short to favor the

main reaction against any interfering reactions. Interest in hydride
generation methods among clinical chemists has always been slight,
but FIA provides a significant step toward greater automation of
the technique and may lead to renewed interest.

5.3 Graphite Furnace/Electrothermal Atomization

The determination of trace metals using electrothermal atomization in
the graphite furnace is often performed after only a predilution
(i.e., lead or cadmium). For many other metals the matrix of the
sample interferes with the assay and an extraction is required to
eliminate the interference. A group in Sweden (60) have extended
the work of Nord and Karlberg (57) on extraction methods for flame
atomization to graphite furnace techniques. They have developed a
two-stage extraction method for the concentration and extraction of
metals (cadmium, copper, iron, lead, and zinc) from aqueous solutions. The extracted samples can then be placed in suitable vessels
for introduction by an autosampler into a graphite furnace. This
combination technique might allow methods that suffer badly from
matrix interference, such as blood aluminium measurements, to be
improved.

5.4 Chromatography/Atomic Absorption Spectrometry

5.4.1 High-Performance Liquid Chromatography/AAS

Coupling high-performance liquid chromatography (HPLC) to an
atomic absorption spectrometer opens many interesting avenues of
research for the clinical chemist. The combination of a separation
technique with a specific metal detector affords the opportunity to
investigate many metal-ligand interactions that occur in the human
body. An initial application in this area has been the study of
copper and zinc in serum (61). Renoe and co-workers (62) have
introduced a flow injection instrument between the outlet of the
HPLC and the nebulizer of the AAS. This injected 11 µL of the
HPLC column eluate into the nebulizer every 10 s producing a series
of FIA peaks corresponding to the concentration of element (calcium
or magnesium) in that part of the eluate. Observing these peaks
over a period of minutes showed larger peaks superimposed, corresponding to the retention times of albumin or citrate-complexed element. Thus for calcium the HPLC-FIA-AAS combination permitted
the simultaneous assessment of the albumin-bound, citrate-complexed,
and free calcium.

The ability to measure metal speciation and/or metal-ligand combinations is of emerging importance in the clinical field and should
be applicable to a wide variety of situations, including metallic poisoning (diagnosis and monitoring) (61) and protein-metal interactions

(63). The use of an FIA instrument as an interface between the HPLC and AAS instruments allows these two components to be optimized separately and connected via the FIA at a later stage. The greater flexibility introduced in this way should enhance the combination of the two techniques.

5.4.2 Ion-Exchange Chromatography/AAS

The lack of sensitivity of flame atomic absorption methods for some metals which are of clinical interest, notably lead and cadmium, have caused most workers to resort to electrothermal atomization or related nonflame techniques for the assay of these metals. An alternative approach is to preconcentrate the metal. One method is to use a solvent extraction-concentration method, as described previously (60). Alternatively, a chromatographic column can be used to concentrate the metal of interest. Hirata et al. (64) have employed a miniature ion-exchange column of Chelex-100 to increase the sensitivity of flame atomic absorption spectrometry for cadmium 15-fold. Samples (4 mL) were acidified (pH 2 to 5), mixed with ammonium acetate buffer (pH 7), and pumped through the column using one pump, the cadmium being retained on the column. When the whole of the sample had gone through the column, the first pump was stopped; a valve changed the input of the column to a second reagent line, through which a separate pump was propelling 2 M nitric acid. This eluted the cadmium from the column and the eluant was fed directly into the nebulizer of the atomic absorption instrument. A precision of 1.7% CV was obtained with aqueous standards at a throughput of 24 samples per hour, and the method was applied to a biological standard reference material (Pepperbush, NIES No. 1). This method could theoretically be applied to the assay of metals, such as aluminum as well as cadmium.

5.5 Inductively Coupled Plasma Emission Spectrometry

Flow injection methods have also been used as a means of sample introduction into inductively coupled plasma atomic emission spectrometers. This facilitates the simultaneous determination of several clinically relevant elements from a single injection (65,66). For multielement determinations the inductively coupled plasma (ICP) consumes less sample and is faster than flame AAS. For the clinically important elements, the detection limits of ICP and flame AAS are similar. However, since ICP instrumentation is about five times as expensive to purchase as flame AAS equipment, it is rarely found in hospital laboratories.

6 CONCLUSIONS

Flow injection analysis is a flexible and convenient technique for sample introduction in atomic absorption spectrometry. Samples can be pretreated in a variety of ways, ranging from simple dilution to solvent extraction. Consumption of both sample and reagent can be reduced to a minimum with carefully designed manifolds and injection methods. Controlled dispersion analysis eliminates the necessity to waste sample, either in filling sample loops or in washing out previous samples, resulting in an extremely low sample requirement. The latter feature is particularly attractive to the clinical chemist. It is surprising, therefore, that there has been relatively little interest to date in FIA in the field of clinical chemistry. This may be related to the absence of a major manufacturer from the FIA-AAS marketplace. The production by such a company of an FIA attachment for their instruments could stimulate an increase in the clinical chemistry applications. FIA-AAS should be a useful and valuable technique in the clinical chemists repertoire, but growth is slow at present.

REFERENCES

1. A. Taylor, Usefulness of measurements of trace elements in hair, Ann. Clin. Biochem., 23:364 (1986).

2. J. Růžička and E. H. Hansen, Flow injection analysis. Part 1. A new concept of fast continuous flow analysis, Anal. Chim. Acta, 78:145 (1975).

3. K. K. Stewart, G. R. Beecher, and F. E. Hare, Rapid analysis of discrete samples: the use of non-segmented continuous flow, Anal. Biochem., 70:167 (1976).

4. B. F. Rocks and C. Riley, Flow-injection analysis: a new approach to quantitative measurements in clinical chemistry, Clin. Chem., 28:409 (1982).

5. B. F. Rocks, R. A. Sherwood, and C. Riley, Flow injection analysis in clinical chemistry, Talanta, 31:879 (1984).

6. E. A. G. Zagatto, F. J. Krug, H. Bergamin F⁰., S. S. Jørgensen, and B. F. Reis, Merging zones in FIA. Part 2. Determination of calcium, magnesium and potassium in plant material by continuous flow injection atomic absorption and flame emission spectrometry, Anal. Chim. Acta, 104:279 (1979).

7. J. F. Tyson, Flow injection methods and atomic absorption spectrometry, Anal. Proc., 20:488 (1983).

8. B. F. Rocks, R. A. Sherwood, and C. Riley, Direct determination of calcium and magnesium in serum using flow-injection analysis and atomic absorption spectroscopy, Ann. Clin. Biochem., 21:51 (1984).

9. J. L. Burguera, M. Burguera, M. Gallignani, and O. M. Alarcón, More on flow injection/atomic absorption analysis for electrolytes, *Clin. Chem.*, *29*:568 (1983).

10. J. L. Burguera, M. Burguera, and O. M. Alarcón, Determination of sodium, potassium, calcium, magnesium, iron, copper, and zinc in cerebrospinal fluid by flow injection atomic absorption spectrometry, *J. Anal. At. Spectrom.*, *1*:79 (1986).

11. N. Leon, J. L. Burguera, M. Burguera, and O. M. Alarcón, Determination of cobalt and manganese in blood serum by flow injection analysis and atomic absorption spectroscopy, *Rev. Roum. Chim.*, *31*:353 (1986).

12. B. F. Rocks, R. A. Sherwood, L. M. Bayford, and C. Riley, Zinc and copper determination in microsamples of serum by flow-injection and atomic-absorption spectroscopy, *Ann. Clin. Biochem.*, *19*:338 (1982).

13. M. Burguera, J. L. Burguera, and O. M. Alarcón, Flow injection and microwave-oven sample decomposition for determination of copper, zinc and iron in whole blood by atomic absorption spectroscopy, *Anal. Chim. Acta*, *179*:351 (1986).

14. M. Burguera, J. L. Burguera, P. Cergio Rivas, and O. M. Alarcón, Determination of copper, zinc and iron in parotid saliva by flow injection with flame atomic absorption spectrophotometry, *At. Spectrosc.*, *7*:79 (1986).

15. B. F. Rocks, R. A. Sherwood, Z. J. Turner, and C. Riley, Serum iron and total iron-binding capacity determination by FIA with atomic absorption detection, *Ann. Clin. Biochem.*, *20*:72 (1983).

16. J. L. Burguera, M. Burguera, L. La Cruz, and O. R. Naranjo, Determination of lead in the urine of exposed and unexposed adults by extraction and flow injection/AAS, *Anal. Chim. Acta*, *186*:273 (1986).

17. B. F. Rocks, R. A. Sherwood, and C. Riley, Direct determination of therapeutic concentrations of lithium in serum by flow injection analysis with atomic absorption spectroscopic detection, *Clin. Chem.*, *28*:440 (1982).

18. J. L. Burguera, M. Burguera, and M. Gallignani, Direct determination of sodium and potassium in blood serum by flow injection and atomic absorption spectrophotometry, *Ann. Acad. Bras. Cienc.*, *55*:209 (1983).

19. W. R. Wolf and K. K. Stewart, Automated multiple FIA for flame atomic absorption spectrometry, *Anal. Chem.*, *51*:1201 (1979).

20. A. S. Attiyat and G. D. Christian, Flow injection analysis-atomic absorption determination of serum zinc, *Clin. Chim. Acta*, *137*:151 (1984).

21. K. W. Simonsen, B. Nielsen, A. Jensen, and J. R. Andersen, Direct microcomputer controlled determination of zinc in human serum by flow injection atomic absorption spectrometry, *J. Anal. At. Spectrom.*, *1*:453 (1986).

22. K. K. Stewart, G. R. Beecher, and P. E. Hare, Rapid analysis of discrete samples: the use of nonsegmented continuous flow, *Anal. Biochem.*, *70*:167 (1976).

23. K. M. Hambridge and B. L. Nichols (Eds.), *Zinc and Copper in Clinical Medicine*, SP Medical and Scientific Books, Jamaica, N.Y., 1978.

24. N. W. Solomons, I. H. Rosenberg, H. H. Sandstead, and K. P. Vo-Khactu, Zinc deficiency in Crohn's disease, *Digestion*, *16*:87 (1977).

25. G. T. Strickland and L-L. Leu, Wilson's disease: clinical and laboratory manifestations in 40 patients, *Medicine*, *54*:113 (1975).

26. D. M. Danks, P. E. Campbell, J. M. Gillespie, J. Walker-Smith, and J. Blomfield, Menkes kinky hair syndrome, *Lancet*, *1*:1100 (1972).

27. K. Fukamachi and N. Ishibashi, Flow injection atomic absorption spectrometry with organic solvents, *Anal. Chim. Acta*, *119*:383 (1980).

28. J. L. Burguera, M. Burguera, O. M. Alarcón, and B. Ibarra de Diaz, Determination of iron, copper and zinc in human vitreous humour by flow injection atomic absorption spectrometry, *Acta Cient. Venez.*, in press.

29. J. F. Zilva and P. R. Pannell, *Clinical Chemistry in Diagnosis and Treatment*, Lloyd-Luke, London, p. 399, 1975.

30. J. W. Jefferson and J. H. Greist, *Primer of Lithium Therapy*, Williams & Wilkins, Baltimore, 1977.

31. J. W. Wollen and M. G. Wells, Simple flame photometric assay of serum lithium, *Ann. Clin. Biochem.*, *10*:85 (1973).

32. V. Lehmann, Direct determination of lithium in serum by atomic absorption spectrometry, *Clin. Chim. Acta*, *20*:523 (1968).

33. A. L. Levy and E. M. Katz, Comparison of serum lithium determinations by flame photometry and atomic absorption spectrometry, *Clin. Chem.*, *16*:840 (1970).

34. J. A. Kanis, A. D. Paterson, and R. G. G. Russell, Disorders of calcium and skeletal metabolism, in D. L. Williams and V. Marks (Eds.), *Scientific Foundations of Clinical Biochemistry*, Vol. II, *Biochemistry in Clinical Practice*, Heinemann, London, 1983.

35. W. B. Vernon and J. H. Greist, Magnesium metabolism, in K. G. M. M. Alberti (Ed.), *Recent Advances in Clinical Biochemistry*, Vol. I, Churchill Livingstone, Edinburgh, 1978.

36. J. B. Willis, The determination of metals in blood serum by atomic absorption spectroscopy. 1. Calcium, *Spectrochim. Acta, Part B, 16:*259 (1960).

37. A. Zettner and D. Seligson, Application of atomic absorption spectrophotometry in the determination of calcium in serum, *Clin. Chem., 10:*869 (1964).

38. E. G. Gimblet, A. F. Marney, and R. W. Bonsnes, Determination of calcium and magnesium in serum, urine, diet and stool by atomic absorption spectrophotometry, *Clin. Chem., 13:*204 (1967).

39. J. Pybus, Determination of calcium and magnesium in serum and urine by atomic absorption spectrophotometry, *Clin. Chim. Acta, 23:*309 (1968).

40. D. L. Trudeau and E. F. Freier, Determination of calcium in urine and serum by atomic absorption spectrophotometry, *Clin. Chem., 13:*101 (1967).

41. P. A. Cooke and W. J. Price, Analysis with the SP90 atomic absorption spectrophotometer. The determination of metals in clinical and biological samples, *Spectrovision, 16:*7 (1966).

42. J. Pybus, F. J. Feldman, and G. N. Brown, Measurement of total calcium in serum by atomic absorption spectrophotometry, with use of a strontium internal reference, *Clin. Chem., 16:* 998 (1970).

43. J. P. Cali, G. N. Bowers, and D. S. Young, A reference method for the determination of total calcium in serum, *Clin. Chem., 19:*1208 (1973).

44. B. F. Rocks, R. A. Sherwood, and C. Riley, More on flow injection/atomic absorption analysis for electrolytes, *Clin. Chem., 29:*568 (1983).

45. B. E. Walker, Clinical applications of flame techniques, in J. E. Cantle (Ed.), *Atomic Absorption Spectrometry*, Elsevier, Amsterdam, 1982.

46. G. D. Christian, The biochemistry and analysis of lead, *Adv. Clin. Chem., 18:*289 (1976).

47. I. L. Shuttler and H. T. Delves, Determination of lead in blood by atomic absorption spectrometry with electrothermal atomisation, *Analyst (London), 111:*651 (1986).

48. H. T. Delves, A micro-sampling method for the rapid determination of lead in blood by atomic absorption spectrophotometry, *Analyst (London), 95:*431 (1970).

49. S. Selander and K. Cramer, Interrelationships between lead in blood, lead in urine and ALA in urine during lead work, *Br. J. Ind. Med., 27:*28 (1979).

50. E. J. Underwood, Cobalt, in *Trace Elements in Human and Animal Nutrition*, 4th ed., Academic Press, New York, 1977.

51. E. J. Underwood, Manganese, in *Trace Elements in Human and Animal Nutrition*, 4th ed., Academic Press, New York, 1977.

52. C. Riley, L. H. Aslett, B. F. Rocks, R. A. Sherwood, J. D. McK. Watson, and J. Morgon, Controlled dispersion analysis; flow injection without injection, *Clin. Chem. 29:*332 (1983).

53. R. A. Sherwood, B. F. Rocks, and C. Riley, Controlled-dispersion flow analysis with atomic absorption detection for the determination of clinically relevant elements, *Analyst (London)*, *110:*493 (1985).

54. N. Yoza, Y. Aoyagi, S. Ohashi, and A. Tateda, Flow injection system for atomic absorption spectrometry, *Anal. Chim. Acta*, *111:*163 (1979).

55. B. F. Rocks, S. M. Wartel, R. A. Sherwood, and C. Riley, Determination of albumin with bromocresol purple using controlled dispersion flow analysis, *Analyst (London)*, *110:*669 (1985).

56. H. Bergamin F⁰., E. A. G. Zagatto, F. Krug, and B. F. Reis, Merging zones in FIA. Part 1. Double proportional injector and reagent consumption, *Anal. Chim. Acta*, *101:*17 (1978).

57. L. Nord and B. Karlberg, Sample pre-concentration by continuous flow extraction with a flow injection atomic absorption detection system, *Anal. Chim. Acta*, *145:*151 (1983).

58. J. C. de Andrade, C. Pasquini, N. Baccon, and J. C. van Loon, Cold vapor atomic absorption determination of mercury by flow injection analysis using a Teflon membrane phase separator coupled to the absorption cell, *Spectrochim. Acta, Part B*, *38:*1329 (1983).

59. O. Åström, Flow injection analysis for the determination of bismuth by atomic absorption spectrometry with hydride generation, *Anal. Chem.*, *54:*190 (1982).

60. K. Bäckström, L-G. Danielsson, and L. Nord, Sample workup for graphite furnace atomic absorption spectrometry using continuous flow extraction, *Analyst (London)*, *109:*323 (1984).

61. J. C. van Loon, B. Radziuk, and N. Kahn, Metal speciation using atomic absorption spectroscopy, *At. Absorpt. Newsl.*, *16:*79 (1977).

62. B. W. Renoe, C. E. Shideler, and J. Savory, Use of a flow-injection sample manipulator as an interface between a high-performance liquid chromatograph and an atomic absorption spectrophotometer, *Clin. Chem.*, *27:*1546 (1981).

63. J. W. Foote and H. T. Delves, Distribution of zinc amongst human serum proteins determined by affinity chromatography and atomic absorption spectrophotometry, *Analyst (London)*, *108:*492 (1983).

64. S. Hirata, Y. Umezaki, and M. Ikeda, Determination of cadmium of ppb level by column preconcentration-atomic absorption spectrometry, *J. Flow Inject. Anal.*, *3*: 8 (1986).

65. P. W. Alexander, R. J. Finlayson, L. E. Smyth, and A. Thalib, Rapid flow analysis with inductively coupled plasma atomic-emission spectroscopy using a micro-injection technique, *Analyst (London)*, *107*:1335 (1982).

66. C. W. McLeod, P. J. Worsfold, and A. G. Cox, Simultaneous multi-element analysis of blood serum by flow injection-inductively coupled plasma atomic-emission spectrometry, *Analyst (London)*, *109*:327 (1984).

67. C. Riley, B. F. Rocks, R. A. Sherwood, L. H. Aslett, and P. R. Oldfield, A stopped-flow/flow-injection system for automation of α_2 macroglobulin kinetic studies, *J. Autom. Chem.*, *5*(1):32 (1983).

8

Current Trends

MARCELA BURGUERA and JOSÉ LUIS BURGUERA *Department of Chemistry, Faculty of Sciences, University of Los Andes, Mérida, Venezuela*

GILBERT E. PACEY *Department of Chemistry, Miami University, Oxford, Ohio*

1 INTRODUCTION

1.1 General Discussion

Automated analyzers based on continuous flow techniques have
gained increased popularity in analytical chemistry and are now
widely accepted for routine procedures. However, certain properties of continuous flow systems are inherently different from those
familiar to the average analyst. When a stream of material flows
continually through a long tubing or packed bed in which some
process—a chemical reaction, heat or mass transfer, or simple mixing—takes place, one of the following assumptions is made:

1. We may consider that there is no macroscopic variation in flow
    conditions at different locations along the flow path.
2. The elements of the fluid enter the vessel at the same moment,
    move through it with constant and equal velocity, mix complete-
    ly, and leave the system at the same moment.

The absence of air segmentation and the injection of sample
solution into a continuously flowing stream result in a transient
output signal, which is the most distinctive feature of flow injection
analysis (1). FIA is the first method based on nonsegmented flow
that has been proven capable of competing with the established technique of air-segmented flow. This is due to the simplicity, construction, versatility, cost, precision, high sampling frequency, and short
time between sample injection and readout which are characteristic
of FIA.

The field of atomic spectroscopy is actually not one technique
but three: atomic absorption, atomic emission, and atomic fluorescence. Of these, atomic absorption and atomic emission are the most
widely used, and our discussion will deal entirely with them.

Numerous publications to date suggest that AAS is still the
most popular choice and is firmly established. The explosive growth
and great popularity achieved by this technique, which offers a relatively simple way for the determination of elements at the minor or
trace concentration levels, strongly influence the development of alternative techniques with comparable possibilities. A drawback for
all atomic absorption procedures is that only one element at a time
can be measured. Moreover, the dynamic range of these methods
is limited. However, analytical chemists are continually faced with
demands for the determination of several elements at one time in an
increasing number of samples. Thus inductively coupled plasma
atomic emission spectrometry was a practical and fundamental development. As an analytical tool it has gained popularity. ICP-AES
allows for the simultaneous or rapid sequential assay of up to 48
elements over widely varying concentrations. These determinations

TABLE 1 Atomic Absorption Spectrometry Versus Inductively Coupled Plasma Emission Spectrometry

AAS	ICP
Better detection limit with furnace AA	Refractary elements easily determined
Microsampling	Few chemical interferences
Better precision with flame AA	No ionization interferences
Matrix interferences can be eliminated by addition of reagents and methods of standard additions	Improved detection limits, normally 10 to 100 times more sensitive than AAS
Samples containing relatively large diameter particles can be aspirated	Speed of operation, due mainly to its simultaneous analysis capacity
The nebulizer systems are quite trouble-free, requiring only occasional replacement when corroded	Readout capacities, the ICP output results in μg mL^{-1}
Low cost	Interelement emission interferences can be computer-corrected

are made in aqueous or organic solvents in approximately 1 min, thereby introducing a new dimension, speed, to trace metal determination in analytical chemistry. It might appear that the development of this method has answered all the prayers of the atomic spectroscopists. However, this is not the case.

During the last few years there have been numerous discussions concerning the relative merits of AAS versus ICP-AES. Essentially, the techniques complement each other and are most effective when used in this manner. The advantages of each technique are shown in Table 1. Because of the complementary nature of AAS and ICP-AES, commercial firms have designed instruments capable of performing both techniques in a single unit.

An additional AES technique is direct current plasma (DCP). These systems seem to be favored by geologists and other analysts who cannot afford the cost of ICP. The sequential DCP-AES instruments have many of the same characteristics as ICP-AES and unfortunately, many of the problems. It is reasonable to assume that new FIA approaches that work for ICP-AES would work with DCP-AES.

The recent substantial progress in the multielemental determination capacity of atomic absorption instrumentation seems to provide a solution to the greatest disadvantage of AAS. The key to the success of AAS and the greatest hindrance to the development of multielemental AAS is the light source. In the past nine years a simultaneous multielement atomic absorption continuum source (SIMAAC) spectrometer has been developed that overcomes many of the problems associated with continuum sources by modifying the rest of the instrumental components (2). The success of SIMAAC can be attributed to the synergistic combination of high-resolution echelle polychromator, wavelength modulation, and computerized high-speed data acquisition and processing. Each of these components is essential to the successful acquisition of useful analytical data from a continuum source. Without question the SIMAAC system has a clear potential application for trace metal determinations by AAS.

Despite the fact that some of the earliest FIA papers reported that the technique could be used to present a small volume of sample to a sensor or instrument without prior chemical reaction, and that the advantages of flame microsampling or discrete nebulization were well documented, the first reports of the FIA-AAS and FIA-ICP combinations did not appear until 1979 and 1980, respectively (3-5). The flow design in these pioneering works were based on sample introduction in a dispersing carrier solution which is pumped into the atomic spectroscopic instrument nebulizer. AAS detection is to date the most popular choice in FIA, but there has been an increase in the use of ICP with flowing streams, probably because the simultaneous determination of various elements is feasible, with detection limits in the range 0.01 to 0.2 mg L^{-1} (depending on the element).

In any case, flow injection atomic spectroscopy-based procedures have made a great impact on many areas of chemical analysis. In combination with FIA, these methods improve the sample throughput and simplicity of sample pretreatment, reduce reagent consumption, and make possible total automation of simultaneous analysis. Both combinations have a wide range of applications in busy laboratories where a workhorse system is required.

Owing to the substantial progress that has been achieved in FIA-AS, thereby placing it among the principal analytical tools, it is self-evident that the statement made by Růžička and Hansen in the first book on FIA to be published (1) is still valid for any author who wishes to predict the future development of FIA-AS procedures: Knowledge of the past does not qualify anyone to make sound future projections. Even so, the present authors present here a survey of possible future developments of FIA-atomic absorption.

2 THEORETICAL CONSIDERATIONS

The theoretical models commonly used in FIA have been described by Růžička and Hansen (1). These theories are drawn from a wealth of information and experience from chemical reactor engineering and from analytical chromatography to FIA. Although the development of theoretical expressions for sample dispersion in FIA-AS systems has been approached in several ways, the state of the theory of FIA-AS, unfortunately, is still at a rather lamentable level compared with the sophisticated level of many of the practical developments. The theoretical description of dispersion in FIA related to instrument response in atomic spectroscopy has resisted accurate mathematical solution, and an entirely satisfactory explanation of this variable effect has yet to be given. Because of the complex multivariate processes involved, it is even difficult to describe a single-line system comprehensively.

Despite the dedicated efforts of many workers (6-10), it is not yet possible to give an exact expression for dispersion coefficients in terms of such practical aspects as sample properties, tubing dimensions and material, and operating conditions (8). In particular, much of the current research on FIA-AAS response has given additional insight into aerosol quality and nebulizer performance, quantified by an array of specially defined parameters (8-10). Such information is of great value in improving instrument performance. However, it appears unlikely that absorbance will ever be expressed directly as a function of sample properties, nebulization conditions, and excitation mechanisms, due to the practical differences in many complex intermediate steps. In such situations, simple physical models will be based on simulation and simplification rather than rigorous mathematical treatment, and therefore would allow predictions of systems behavior and the continuation of theoretical developments.

The development of future theoretical expressions should continue to incorporate constants into generalized equations in order to incorporate all the parameters that may affect peak height, such as viscosity and manifold construction. But it should be emphasized that each particular manifold will have its own individual characteristics of total dispersion. Thus a generalized theoretical approach to FIA-AS is likely to provide only a rough guide to the choice of practical conditions. With the ever-increasing number of atomic spectroscopic systems into which the concepts of FIA are incorporated, this task becomes progressively more difficult. The authors of these papers should take into account the peculiarities of each design and the sample properties, and then subject their conclusion to rigorous experimental testing. In any case, care has to be exercised when theoretical conclusions are to be drawn from applied FIA-AS systems.

3 INSTRUMENTAL DEVELOPMENTS

3.1 Individual Components

The simplest, single-line FIA-AS system (Fig. 1) consists of a pump, which is used to propel the carrier stream, R, through a tubular channel; an injection unit, I, by means of which a well-defined reproducible volume of a sample solution is injected by a valve or hydrodynamically into the carrier stream; a coil, C, in which the sample zone disperses or where the analyte may conveniently be diluted or preconcentrated; an AA, ICP, or DCP detector, D; and a fast response chart recorder or high-speed data acquisition device and processor, DA.

3.1.1 Pumps

For many practical assays results have been reported for systems in which a pump was not used (11-13). The driving force was the reduced pressure developed at the nebulizer by virtue of the oxidant flow. It would appear that this procedure is less satisfactory than using a pump, for the following reasons.

The carrier stream flow rate depends on the aspiration rate of the nebulizer, which is accurately controlled by adjustment of the valve regulating airflow to the nebulizer. However, as the aspiration capacity of the spectrometer alone is not enough to provide an adequately constant flow rate for precise results (14), the use of peristaltic pumps should continue to mediate the residence time of the sample in the manifold. In this way, more precise results are obtained (15,16), the system is less sensitive to changes in sample viscosity (15), the degassing and ingress of air at connections is almost eliminated (17), and an appropriate flow rate is maintained even with long lengths of tubing (15). For many applications various types of pumps have been used (18), including peristaltic, single and dual piston, and continuous-pressure pumps (19).

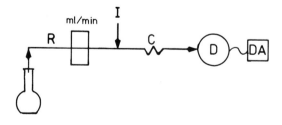

FIGURE 1 Single-line manifold. R, carrier stream; I, sample injection; C, dispersion coil; D, detector; DA, recorder or a data acquisition unit.

Normally, the application dictates the choice of a particular pumping system. For moving corrosive fluids the peristaltic approach is probably the least desirable, although they do provide for the lowest cost per channel when used for pumping aqueous fluids. Multichannel pumps have certain disadvantages, particularly when a new method is under development, because the flow will pulsate slightly as a result of movement of the rollers along the pump tubing. Also, to vary the amount of a given reagent, the diameter of the pump tubing carrying this reagent must be altered. When this is done the continuous flow is normally interrupted for all channels. Varying the speed of the pump is of little use, because this will alter the rate of the sample and reagent in the same way (18). Fortunately, in this respect it should be emphasized that fewer pumping channels are commonly used in FIA-AS systems, and in many cases, instead of using a single multichannel pump, it may be advantageous to employ separate minipumps with one or two channels each. The many advantages of using minipumps is that they are cheaper and offer unique flexibility by having a completely independent working capacity for each channel or pair of channels. It can also be placed near the respective modules, so that long tubing between the different parts of the system can be avoided. This may have a favorable effect on the response time of the system because cross-mixing of sample is minimized.

3.1.2 Sample Injectors

The need to determine samples with high dissolved solids and the variations in viscosity or solvent content may make continuous nebulization difficult, impossible, imprecise, and dangerous (9). Therefore, it is well recognized that all these disadvantages are overcome by introducing a small volume of analyte into the nebulizer using two procedures: discrete nebulization (20,21) and flow injection. Although the discrete nebulization process uses inexpensive injection devices, changes in background, flame geometry, memory effects, and deterioration of precision may occur. The fact that it is difficult to add releasing agents or spectroscopic buffers to a small volume to be injected is also a disadvantage (9,22). Flow injection procedures suffer less from these problems since there is a continuously flowing carrier stream washing the nebulizer.

Numerous designs of sample injectors have been tried, often characterized by increasing levels of complexity. The initial approach of sample introduction with a small syringe in combination with a hypodermic needle through the wall of the tube or through a septum (23) was replaced by a syringe in combination with a flap valve (24). These methods have several defects; for example, the use of syringes is tedious and too slow for practical assays since the needle disrupts the flow. There is considerable mixing so the contribution of the injector process to the dispersion may be considerable. Injection through rubber septa is also difficult to mechanize completely, and nebulizer blockage is often caused by pieces of rubber cut out by the needle.

The majority of workers have favored some form of valve capable of transferring and accurately measuring the volume of sample into the flowing carrier stream. Many manually or motorized types of valve design (24-27) have been incorporated in flow injection systems to perform sample injection between high- and low-pressure lines (28). Most of them are homemade and others are not specifically manufactured for FIA, such as rotary (29), solenoid slider (30), custom built (31), or commutator valve (32). At the top of the range are the electronically controlled injector commutators, developed by a Brazilian group (32), which are capable of performing merging-zone and zone-sampling procedures. For most applications, the various approaches seem to be comparable in capability. The primary considerations when choosing a valve for a particular application are smooth injection, capability of wetted surfaces with sample and stream components, and accommodation of the required injection volume.

As mentioned previously, for many applications steady-state equilibrium is not attained in FIA. Therefore, all these designs have one common denominator: The solution to be injected must be metered exactly and transferred reproducibly into the carrier stream.

All systems employing rotary valves suffer from a common fault: Not only does the sample fill the volume of the sample loop, but the entry and exit tubes must also be filled with sample. Additional sample is used to flush out the residual of the previous specimen. Therefore, the injection process is wasteful of sample. This is a serious disadvantage, because economy of sample should be one of FIA's virtues. Loops may be bent or otherwise damaged, giving erroneous results which necessitate immediate corrective action. Rotary valves are also prone to leakage when exposed to the heavy work load of a routine laboratory.

Alternative valveless approaches to sample introduction in a FIA system have been described by several authors based on controlled dispersion (33), hydrodynamic (34), and nested (35) injection principles. Controlled dispersion analysis is similar to Růžička and Hansen hydrodynamic injection (34), which offers zero dead volumes, but is slightly less elegant in that it contains moving parts. However, hydrodynamic injection still requires that connecting tubing be filled, and the volume injected is not easily varied. The performance of both systems appears to be comparable to that of a valve.

Although some of the injection systems mentioned above have not yet been used with FIA-AS manifold, there is a marked tendency

to use the principles of injection commutators (32) in fully automated processing systems for the handling of solutions, including its
combination with the hydrodynamic, nested, and simultaneous injection procedures. Also, controlled dispersion analysis (33) and single standard calibration (36,37) schemes will be continuously used
to introduce sample volumes into the carrier stream more efficiently.
For the FIA-AS to be used as a workhorse in a busy analytical laboratory, the development of intelligent automatic samplers is most
desirable. Since flow injection can perform as a high-speed analytical system, the analytical apparatus should be capable of accommo-
dating relatively large batch sizes; this minimizes the operator attention required to change sample trays. The successful adaptation of
electronically controlled injector commutators to FIA-AS systems
should not only economize sample and reagents, but will be capable
of performing merging-zone and zone-sampling procedures.

3.1.3 Detectors

Detectors for FIA are usually flow-through types, designed to cause
minimum disturbance of the flowing stream so that detector contribution to background signal is negligible. An ideal detector in flow
injection systems must have a fast response to cope with the high
sampling rates that can be achieved. It should respond linearly
over a wide concentration range of the components transported to
the detector. Responses should be repeatable over extended periods
of time so that frequent calibration runs necessitated by detector
weakness are not required (38). The detector should have very
low inherent electrical noise characteristics and be sensitive enough
to yield the highest possible reproducibility, as well as being compatible with data processing systems. To achieve these requirements, special attention must be given to manifold design and construction. Thus the output signal generated by the detector should
conveniently be displayed with a fast-response chart recorder as
sharp peaks or processed by a computerized high-speed data acquisition. Used as a FIA detector, the atomic absorption and ICP spectrometers have some limitations. In both cases, the sample flow is
totally disrupted during nebulization, to create an aerosol suitable
for atomization. The resulting analytical signal relates to peak concentration of aerosol entering the flame or plasma rather than to the
traditional FIA transient concentration. The process of aerosol generation and conditioning takes time, so that response is not instantaneous. The performance of such systems are not as good as
those obtained with continuous nebulization. The sensitivity is less
and the precision is poorer or the same. Although it is difficult to
envisage how this could be otherwise, it is possible to give the
following guidelines:

1. The precision can be improved by using a pump to deliver solution at a flow rate slightly above the instrument's conventional optimized rate.
2. Measure peak area instead of peak height.
3. Improve nebulization efficiency.

Therefore, an ideal nebulizer should

Aspirate sample at an extremely steady rate for long periods of time
Be easy to adjust, assemble, and disassemble
Produce uniformly fine droplets
Be inert to chemical attack by a wide range of acids and organic
    solvents
Handle a wide viscosity range with respect to sample solutions
Be relatively free from potential plugging and easy to clean
Have good washout characteristics to minimize memory effects

Further studies and improvements in the nebulizer design should be directed toward increasing the residence time of analyte species either in the flame or plasma and reducing possible solvent interactions in the atomization process. Both approaches should improve the limits of detection.

3.2 Variety of Manifold Designs

The final design of a flow injection system is dependent on the particular requirements of the user. However, critical to the successful operation of a flow injection analyzer is thoughtful design of the analytical manifold. Since sample dispersion is being used to effect specific functions, including sample dilution, mixing, and reaction, it should be controlled as meticulously as possible. Knowing the parameters affecting the dispersion of an injected sample in a FIA-AS system makes the design of manifolds for specific application feasible. As a given FIA-AS manifold may consist not only of the sample treatment components (holding coils, extraction apparatus, ion-exchange columns, etc.), fittings used to interconnect the various components and modules should be well made, to minimize dead volumes.

Further development and refinement of FIA manifolds in combination with AAS and ICP will be developed for multielement preconcentration, removal of interfering constituents from the sample speciation, and conversion methods (39), with a high sampling frequency. Therefore, a variety of different kinds of manifolds will be contemplated to improve detection for FIA-AS using either solvent extraction, ion exchange, or kinetic chemical exchange. Thus the simultaneous multielement capabilities of ICP-AES and SIMAAC combined with the substantial preconcentration and/or removal of matrix effects for on-line FIA systems would continue to open new possibilities for analyte determinations at ultratrace levels. Also, it is conceivable that the present systems could be operated at even higher sampling rates. The only reason that this possibility was probably not tested was the nonavailability of a dedicated microprocessor designed for these processes.

In seeking to exploit the chemistry fully, we need a FIA-AS system combined with several other kinds of detectors, such as electrochemical and spectrophotometric detectors. This conceivable combination will allow the analysts to handle the input from the several detectors, improve the measurement of individual species, determine their chemical forms, and extract quantitative data by means that at present are unconventional.

3.3 Developing Methods for Sample Preparation

The modern analytical laboratory is frequently called upon to perform multielement and simultaneous multielement analysis of widely diverse sample types. Therefore, one of the major challenges in analytical chemistry and spectroscopy has been to develop an analytical system that would allow the determination of all the elements, from the ultratrace to the major constituent level, in an acceptable manner. It is doubtful that a single analytical instrument or technique will ever fulfill all the requirements for such an ideal system. Particularly, atomic spectroscopic instruments usually offer a high rate of sample handling, and in many applications this rate is not easily matched by the preparation steps needed to obtain a solution suitable for analysis. Therefore, the development of complementary techniques which can greatly simplify some of the common preparation steps and which should be readily used with any spectrometer is needed. An ideal pretreatment system for atomic spectroscopic analysis should offer the following benefits:

Converting sample form to one acceptable for analysis
Obtaining representative samples
Elimination of matrix component introduced into the nebulizer which
would otherwise cause blockage or damage
Sample cleanup with the elimination of spectral or chemical interferences from the matrix
Sample dilution to place high-analyte-containing samples into a measurable range
Sample preconcentration, to increase sensitivity to suitable levels
Controlling contamination and losses

The requirements above are fulfilled to an unusually high degree by FIA. One of the major benefits of FIA compared with other types of automated analysis systems is its capability of performing

sample pretreatment functions in-line with great flexibility, economy
of sample, and reagents, due to the elimination of large amounts of
manual work. These techniques serve primarily to remove interfering substances from the sample matrix or to place the analyte into
a detectable form. The most important of these treatments include
those which can be used more or less routinely and in fully automated form in a FIA system

Dilution of the sample
Dialysis (liquid and gas)
Ion exchange
Oxidation or reduction
Liquid-liquid extraction
Addition of common reagents
Filtration
Hydride generation systems

Preconcentration of samples is usually performed by either
liquid-liquid extraction or ion-exchange columns. While solvent extraction techniques have been widely extended in FIA-atomic spectroscopy, this method is far from ideal because it requires careful
work and its blank values are to be kept low. Further, analysis
of a large number of sample yields, along with the analytical results, considerable volumes of used organic solvents that have to
be disposed of in an environmentally acceptable way. Therefore,
there has been a tendency to use another preconcentration technique, such as ion-exchange chromatography. However, many of
these objections have been reduced by using microextraction systems. Ion-exchange techniques are valuable systems that have
been used successfully for various matrix modifications to guaran-
tee reliable results for samples such as seawater, river water, and
groundwater. In this sense, novel chelating resins should be syn-
thesized and applied with FIA-AAS and FIA-ICP as a means of dis-
tinguishing between labile and inert forms of metals and to enhance
the practical sensitivity for ultratrace determinations in biological
and other materials. It is important to point out that the size of
the column used to date is in the range of 2 cm long with a 0.5-cm
inside diameter. Backflushing and reconditioning steps can be accomplished between by placing the column in place of the sample
loop of another injection. Table 2 shows the advantages of using
liquid-liquid extraction and ion-exchange procedures in combination
with a FIA-AS system. Thus FIA based on liquid-liquid extraction
and ion-exchange techniques has the potential to open up even more
new fields of practical assays.
    The recent development in automated hydride generation has
resulted in better systems with regard to precision, sample

TABLE 2   Advantages of Using Liquid-Liquid Extraction and Ion-Exchange Techniques in Combination with FIA for Atomic Spectrometric Determinations

Liquid-liquid extraction and ion-exchange techniques	Additional advantages of ion-exchange techniques
Low sample consumption	Greater concentration enhancements (from 10 to 500 times)
Fast determinations	Greater selectivity, especially for studies of rare-earth elements
Simple assemblies	Continuous control of the column, which makes possible speciation of the trace elements under study
Low cost	Long-term stability of the column packing material under acidic and basic conditions
Versatile	Sharper peak profiles obtained on elution

throughput, and simplicity of operation. However, the hydride generation technique still suffers from many disadvantages, most of which seem to originate from the poor control over the factors involved in the hydride generation step (40). The recently reported dual-phase gas diffusion system for hydride generation completely eliminated the transition metal interferences (41). The FIA technique appears to be very promising for controlling interference effects when used in hydride generation systems with atomic spectrometric detection. Among other qualities of such a combination are unique possibilities of miniaturization, great flexibility, and useful for gaining fundamental knowledge about the chemical reactions involved (42).

Because the methods of analysis described here require sample material to be in solution, solubilization of the solid sample is necessary. This is generally performed after drying or dissolving the various samples. Thus the most time-consuming step in the procedure of any method is the dissolution of the sample. Thus there is an urgent need for developing rapid preparation methods to dissolve powdered and solid samples in a fully automated manner. This need has led several workers to investigate the use of a microwave oven as a heat source in a low-temperature digestion system for acid dissolution of metal, geological, and biological materials with favorable results (43-45). This method has proved to be rapid, inexpensive, and highly compatible with FIA-AAS for the acid treatment of whole blood (46). However, there is a great challenge for future investigations to develop on-line systems for solid samples preparation in FIA-AS.

3.4   Multielement and Simultaneous Multielement
Determinations

It is important to distinguish between multielement and simultaneous
multielement determinations.   The idea of multielement analysis is the
possibility of determining two or more elements by a given instrumental system, generally using parameters highly specific to each
element being determined (47,48).   Frequently, the parameters selected are not only element specific but also sample specific.   Although simultaneous multielement determinations are feasible using
a given instrumental system, these determinations to require the
selection of a single set of "compromise" parameters.   Ideally, the
compromise parameters will not result in a decrease in the accuracy
or the detection limits of any of the elements.   The selection of
compromise analytical parameters in an atomic spectrometric system
goes beyond the selection of atomization parameters and may include
the sample preparation, sample collection, and sampling pattern (47).

The idea of simultaneous multielement analysis by AAS has not
found general application, largely because of the limits imposed by
the absorption law itself.   The concentration range over which an
element can be determined with acceptable accuracy in a given solution is comparatively small.   It is unlikely, therefore, that when two
or more elements are to be determined simultaneously in the same
sample solution, their concentrations will all occur in the correct
range (49).   However, the recent development of SIMAAC (2) offers
some unique analytical features to the field of AAS.   With a simple
optical configuration, SIMAAC achieves simultaneous multielement detection, background correction, large calibration ranges, and detection limits comparable to line-source AAS above 280 nm.   The strength
of SIMAAC lies in the computer programs that control its operation
and process multielement data at extremely rapid rates.   Thus a combination of FIA-SIMAAC has led to the simultaneous flame AA analysis of up to 16 elements per sample at the rate of 60 samples per
hour (50); the ability of this system to modulate over a wide wavelength interval allowed the use of extended dynamic range techniques
to obtain linear dynamic ranges of four to five orders of magnitude.

Without any question, ICP at its current stage is the analytical
technique that provides the most adequate simultaneous multielement
analysis capability.   ICP allows the simultaneous assay of several
elements at major, minor, trace, and ultratrace levels without changing any experimental parameter in aqueous or organic solvents.

The analytical information obtained by any of the AS techniques
can be extended when they are combined with FIA.   With FIA it is
possible to obtain sequential values of the analytical signal at different times or with different instrumental variables, and to obtain
values simultaneously by a single detector.   This can be done in
two different ways

1. By use of several detectors of different nature arranged in series or in parallel, with or without splitting the sample channel, the last detector being an atomic spectrometer.
2. By use of a single detector based either on the establishment of a concentration gradient or on the splitting up of the sample channel into two reactors, which merge in front of the detector. In the last case, the splitting of the sample is due in order to delay the entrance of one of the sample plugs into the detector.

FIA-AS has devoted great attention to multiple detection, generally focusing on the resolution of mixtures of different species. Both series (51,52) and parallel (53,54) techniques have been used for this purpose. However, the ever-increasing necessity to measure several parameters rapidly in the same sample in several practical areas will continue to urge the development of FIA-AS automated methods which could offer the possibility of carrying out simultaneous determinations. Therefore, we are convinced that a broader development of this area, relatively unexplored in FIA-AS, will stimulate the development of commercially available instruments, with particular emphasis on the application of clinical, environmental, pollution, industrial, and agricultural analysis, where the need to know several parameters is very frequent due to the small quantity of sample available.

3.5 Speciation

For the purpose of the argument on speciation it is easier to imagine a plug of sample so that all the chemical forms of the mixture can be distinguished and determined quantitatively. The speciation of elements is of great interest, especially to environmental and clinical analysis, taking into account that the toxicity of an element depends on its clinical form. There is a need for considerable effort to be used to develop analytical methods that distinguish between the different species of a substance which can be found in the sample under investigation. The metal speciation information should be applicable to a wide variety of clinical situations, including the essential status of metal species in normal persons or in cases of metal poisoning.

There are three approaches in FIA which will distinguish between chemical forms of an element: multidetector FIA, separation/single detector FIA (39), and kinetic discrimination (55). It can be visualized that both approaches will be eventually performed on speciation of inorganic and organic forms in diverse complex matrices. Particularly, through the application of novel chelating resins in a FIA system, many of the present restrictions of AAS and ICP for biological, environmental, and agricultural materials will be minimized.

3.6 Conversion

When a new approach is being introduced in FIA, it appears to acquire, quite quickly, its own particular development. Initially, in
some cases, it may appear to be the simple application of previously
known ideas. But the subsequent practical work gives you a better
knowledge of new physiochemical and analytical characteristics of
this particular system. Then the analyst produces a list of stereo-
typed questions relating to such aspects as specificity, sensitivity,
speed, interelement effects, and blank determinations. The basic
principles underlying conversion analysis, in which a nondetectable
species can be converted through a heterogeneous chemical reaction
into a detectable species, is quite a new approach and applied successfully for the indirect determination of cyanide at ppb levels (39,
56). However, the analytical potential for conversion combined with
FIA-AS techniques have not yet been exploited. Forthcoming papers
will surely provide many commendable features and useful applications, especially for trace amounts of certain elements, which have
their atomic resonance lines in the far-ultraviolet region.

3.7 Automation

The ever-increasing demand for analysis in clinical, agricultural,
pharmaceutical, industrial, and other types of analytical control has
led to the development of a large number of different instruments
for automated analysis (57). Developments in this field have been
further stimulated by the additional advantages of automation, such
as increased precision, decreased cost of individual assay, and the
satisfactory reliability of automated equipment. It has been estimated that the market for automated instruments for wet chemistry
achieved an annual growth rate in excess of 15% up to 1980, compared with the growth rate of 9% for all types of analytical instrumentation. The demand for the type, complexity, multiplicity, and
rate of analysis to be performed is very diversified; consequently,
the major instrument manufacturers tend to concentrate on specific
areas such as clinical analysis, which offer a large market and
relatively similar types of material to be analyzed (23).

The equipment of many laboratories often includes continuous
automatic analysis systems of an older generation. This generally
means that sample rate and sample-to-wash ratio are controlled by
cams. Apart from their price, these electromechanical devices show
the disadvantages of poor flexibility and limited sample rate and reproducibility (58). Thus with several hundred samples to be assayed,
the last task has been the further automation of each system. Even
so, the limitation to high rates of sample throughput with continuous
flow automated analyzers is due to an unacceptably high sample carryover which makes any meaningful manual or visual interpretation

impossible. Generally, programs for carryover correction involve relatively large computer capacities.

FIA provides a simple means for complete automation, and the efficiency of this equipment will currently allow analysis at a very moderate cost. In this advanced age of electronics, diverse types of microprocessors can be interfaced economically. These can be used to increase the ease of the system by generating interactive messages for operating program input and to control the functioning of the system using the operator-specified parameters. Furthermore, data reduction can be fully automated using microprocessor technology. Such functions as peak picking, linearization, integration, baseline, drift correction, automatic regulation of sheathing flow rate and nebulization, injection flow rate of the sample to the spectrometer, and function monitoring can also be performed automatically and the results printed out in an organized way by means of a computed simple disk.

The computer programs for quantitative analysis by FIA-AS should quantify the net analyte signal from the measurements in the relevant spectral region, whereby all possible contributions from the background and/or interfering lines are considered. This artificial intelligence should obviate all operator decisions with respect to background correction and spectral interference correction. The programs should enable accurate analysis of samples without prior knowledge of sample composition when simultaneous multielement determinations are carried on, and should require the most minimum operator-software interaction. Thus it can be visualized that automated FIA-AS will eventually be based on

1. *Microprocessor technology development.* In this way the versatility of the instrumentation will increase. However, it should be noted that the cost of such systems may increase considerably. It is therefore worthwhile to pursue the development of instrumentation components at relatively low prices, allowing access to FIA-AS to those who cannot afford to buy sophisticated instruments.
2. *New designed modular equipment.* This modular approach would give maximum flexibility for incorporation of further improvements in any component, and for best use of suitable commercial equipment already available in many analytical and clinical laboratories.

3.8 Miniaturized FIA Systems

The analytical advantages of miniaturized systems with respect to smaller absolute detection limits and speed of analysis are of such great importance and general significance that further extensive

work in this direction appears to be indicated. There are many
convincing reasons for developing miniaturized FIA systems for
atomic spectroscopic detection devices, which would allow lower flow
rates than are usually present. Four obvious advantages are

1. The smaller sample amount needed for the smaller systems
2. The low consumption of solvents and reagents
3. Higher rates of sample throughput
4. Cheaper manifolds

    These miniaturized configurations should allow simultaneous de-
termination of various analytes in a sample either in a sequential
way using sequential detectors or by reversed FIA (59) using a
single detector or splitting based of the flow using sensitive and
selective methods will be described with the incorporation of ion
exchange using resin microcolumns for concentration and/or separation purposes. In addition, several new liquid-liquid separator
devices will be tested and coupled to a FIA-AS system.
    The major drawbacks to be found in a miniaturized system may be

1. A separator is required which, in some cases, should operate
    at high flow rates.
2. The sample dispersion, and therefore its dilution, should be
    maintained beneath an acceptable upper limit as required. The
    development of FIA manifolds capable of producing high sample
    dispersion are frequently requested in clinical analysis. Thus,
    by means of sample predilution, the injection of sample volumes
    of less than 1 µL or the use of small mixing chambers must be
    required.

4 APPLICATIONS

4.1 General Practical Considerations

The diverse nature of the samples encountered, as well as the specific methodologies dictated by regulation bodies in many cases, require that FIA-AS methods be quite versatile. As described previously, there is a vast literature set which explores the application
of FIA-AS to a wide variety of research, industrial, and clinical
problems and describes the automation of virtually every operation
previously used in manual analysis. Undoubtedly, FIA-AS will continue to be a very widely used technique in many areas, such as
agricultural, clinical, and environmental. In addition, we are beginning to see considerable utilization of this technique centered on
analysis for several elements in pharmaceutical, metallurgical, geological, and food mixtures, and the screening of biological particles
such as bacteria and fungi in water and organic materials.

4.2 Clinical Chemistry

Living organisms represent a complex and high organized structure consisting of individual chemical compounds performing highly specialized roles. Therefore, taking into account the number of determinations per day carried out at present by FIA-AS methods, clinical chemistry is undoubtedly its most relevant field of application. But although FIA-AS has been received enthusiastically by analytical chemists working in industry, it has not found immediate favor with clinical chemists. There are four important reasons for this

1. Generally, clinical chemists ignore the enormous potential of the method. They have become used to purchasing dedicated instruments with methodology fully developed and comprehensive data handling systems included. So far no manufacturer has offered such a package based on FIA.
2. The sampling values universally used are wasteful of sample because in order to remove the residue of the previous specimen, an excess of sample must be drawn through the valve and connecting tubes. Besides, such injection valves develop leaks when subjected to the heavy work load of a hospital laboratory.
3. Several frequently requested clinical determinations require systems capable of producing very high dispersion.
4. The FIA-AS apparatus does not possess the sophisticated sample handling and data processing facilities that clinical chemists have come to expect, and the pressures on the clinical chemist are such that they no longer have the time to adapt unsuitable machines.

At present, the major advantages of FIA-AS for clinical analysis are

Rapid production of results
Low reagent consumption
Mechanical simplicity
Short startup and close-down times, making this technique suitable
   for standby or emergency applications

The drawbacks of the FIA-AS techniques given previously can easily be overcome by manufacturing automatic commercial instruments specifically designed for clinical work, with comprehensive data handling systems included. Thus additional advantages can be introduced, such as

The introduction of new types of devices for selective membranes
   and/or chelating resins for specific applications
A wide development of simultaneous determination of several species
   in a particular sample

TABLE 3 Other Applications of Flow Injection Analysis/Atomic Spectroscopy

Element or specie	Detection	Sample medium	Approximate range (mg L^{-1})	Sample volume (µL)	Approximate sampling frequency (h^{-1})	RSD (%)	Ref.
Antimony (Sb)	FIA-AAS	Steel	0–10	90	120	2.9	60
Arsenic (As)	FIA-batch system- HG-ET AAS	River water	0.002–0.008	25–50 mL	15	5	61
	FIA batch system- HG-ET AAS	Biological materials (orchard leaves and oyster tissue)			15	5	61
	FIA-AAS	Food digests[a]	0.12–3.0	300	120–180	2.1	4
	FIA-ICP	Glycerine	0.12–3.0	374		1.4	62
	FIA-HG-ICP	Several materials (orchard leaves, coal fly ash, and river sediment)	0.01–1.0	170	200	7.2	63
Bismuth (Bi)	FIA-AAS	Steels	0–10	90	120	2.9	60
	FIA-HG-AAS	Standard solutions	1–100 ng	700	90–180	0.2–0.8	64
Cadmium (Cd)	FIA-AAS	Foodstuffs	0.04–0.30	30	40	1.9	65
Cerium (Ce)	FIA-AAS	Standard solutions	0.07–0.70	54	50	3.3	66
Chromium (Cr)	FIA-AAS	Steels	0–20	50–100	80	0.48	67
Copper (Cu)	FIA-ICP	Alloys		500	6	2.5	68
	FIA-AAS	Food digests (beans, wheat, and beef)		300	120–180	2.1	4
	FIA-AAS	Powdered milks	0.05–1.00	100	100	5.1	43

Iron (Fe)	FIA-AAS	Powdered milks	0.04-1.00	100	100	5.1	43
Lanthanum (La)	FIA-AAS	Standard solutions	0.02-0.22	54	50	1.8	66
Lead (Pb)	FIA-AAS[b]	Gasoline	0-16	100	300	1.0	69
	FIA-AAS	Foodstuffs	0.10-2.00	30	40	1.9	65
	FIA-AAS	Steels	0-10	90	120	2.9	60
Niquel (Ni)	FIA-ICP	Alloys		500	6	2.5	68
Nitrate (NO_3^-)	FIA-liquid-liquid extraction	Standard solutions	0.1-2.2	130	30-40	3.7	70
Nitrite (NO_2^-)	FIA-liquid-liquid extraction	Standard solutions	0.5-10.0	130	30-40	4.7	70
Selenium (Se)	FIA-batch system-HG-ET AAS	Biological materials (oyster tissue and wheat flour)			15	5	61
	FIA-batch system-HG-ET AAS	River water	0.002-0.008	250-500 mL	25	5	61
Silicon (Si)	FIA-ICP	Phosphoric acid sample	0.4-400	40	20	2.9	71
Silver (Ag)	FIA-AAS	Steels	0-1.0	90	120	2.9	60
Uranium (U)	FIA-AAS	Standard solutions	0.05-1.0	54	50	0.46	72
Zinc (Zn)	FIA-ICP	Alloys		500	6	9.3	68
	FIA-ICP	Food digests[a]		300	120-180	2.1	4

[a]Food digests (beans, wheat, and beef) and biological materials (spinach, orchard leaves, and bovine liver).

[b]Carrier propelled by gas pressure.

As considerable research effort is now in progress on the development of FIA-AS methodology, we are beginning to see major
new trends that will consolidate it completely in the area of clinical
analysis.

4.3 Agricultural and Similar Areas

Several workers have demonstrated that FIA in conjunction with
atomic spectrometric methods is very beneficial for routine determinations of metal ions in waters, soil extracts, and plant materials.
Numerous studies continue to be published in these areas, reflect-
ing the importance of this combined technique for the determination
of various types of chemical species commonly encountered in admix-
ture in samples.   Despite its potential, FIA-AS does not seem to
have been fully exploited.

Attempts to access soil fertility, wastewater, and plant material
composition have used a variety of wet digestion procedures with
acids for mineralization and reagents to extract from these matrices
any element of a particular interest.   Although the use of AAS and
ICP both simplifies and expedites the analysis of the extracts, some
determinations are hindered by the tendency of the nebulizer to
"clay" when solutions containing high salt concentrations or salts of
low solubility are aspirated.   This leads to inaccurate results as well
as delays while the clogging is being removed.   If clogging is to be
avoided, it is necessary to perform additional work on the extract,
such as dilution or separation of the analyte from the interfering
salt.   FIA will continue to offer useful solutions.   Thus simple and
stable FIA systems with the incorporation of ion exchangers and a
liquid-liquid extractor will continue to be used to improve the direct
determination of a wide variety of chemical species in the agricultural and similar areas.

4.4 Environmental

The reasons for wanting to control pollution of our environment are
well founded and well known.   Of the variety of species spread as
a result of human activities, the heavy metals are among the most
dangerous because of their long-lasting effects.

The demands for fast techniques in the control of environmental pollution are rising at an ever-increasing rate.   FIA-AS provides
simple, handy, and versatile instrumentation which makes it suitable
to meet the requirements of any laboratory wanting to make this
type of analysis.   From the birth of this technique to now, relatively few applications to solve environmental problems have been developed.   To date, applications have been mainly on pollutants in
waters and in a more restricted way to air pollutants.   Although no
major trends are noticeable, with its wide field of possibilities and

with the help of analysts' ingenuity, FIA-AS will aid in the solution of many problems in this field. It is necessary to emphasize its applicability to simultaneous analysis through its multiply flexible modifications. Particularly, there will be a major interest in the determination of "hydride-forming" elements and mercury by developing volatilization and stabilization techniques combined with FIA-AS. The extension of this technique to other environmental and geological materials appears promising.

4.5 Other Applications

As can be seen from the previous chapters, the most important applications of FIA-AS were developed for agricultural and environmental materials and body fluids and tissues. However, there are also most valuable applications in metallurgical, food, biological, and other kinds of samples, which are detailed in Table 3. All these applications are based mainly on the association of AAS with FIA. Even so, in our opinion, due to the simultaneous determination of several analytes by ICP and SIMAAC, their association with FIA will also rapidly increase to offer the most diverse solutions to the different shortcomings encountered in the various fields of the analyst interest. At first sight it may appear that all elements in any type of sample are capable of being determined by FIA-AS. However, many limiting factors may be encountered by analysts, which could probably be solved by their ability to apply one of the many features of FIA-AS combinations.

5 FINAL COMMENTS: THE PRESENT/THE FUTURE

In summary, at present the role of FIA-AS in the scientific community is broadly defined. It is appropriate for both laboratory and on-line applications. Furthermore, as we have seen throughout this book, FIA-AS is extremely diverse. Not only does it serve as a dependable workhorse system, but it can also automate different types of analysis, function as a research tool, and give prompt results. Simultaneous multielement analysis of trace elements is feasible with FIA-ICP for just a few sample volumes. The exponential increase in FIA-AS publications since 1979 testifies to some of the innovative possibilities that exist for the technique, and this trend should continue for a number of years. Therefore, in the future, FIA-AS techniques should increasingly offer the advantages of low sample consumption together with a higher frequency of sample analysis per unit time due to fully automated manifolds to be developed that will allow a variety of measurements. The kinetic aspects of the on-line liquid-liquid preconcentration techniques, as well as the efficiency of ion-exchange columns usually used for the elimination of matrix components and sensitivity enhancements, require further study. Some questions remain to be answered; however, it is hoped that in the next phase the importance of FIA-AS, particularly in routine analysis, will increase.

REFERENCES

1. J. Růžička and E. H. Hansen, *Flow Injection Analysis*, Wiley, New York, 1981.

2. J. M. Harnly, Multielement atomic absorption with a continuum source, *Anal. Chem.*, *58*:933A (1986).

3. E. A. G. Zagatto, F. J. Krug, F. H. Bergamin F⁰., S. A. Jørgensen, and B. F. Reis, Merging zones in FIA. Part 2. Determination of calcium, magnesium and potassium in plant material by continuous flow injection atomic absorption and flame emission spectrometry, *Anal. Chim. Acta*, *104*:279 (1979).

4. W. R. Wolf and K. K. Stewart, Automated multiple flow injection analysis for flame atomic absorption spectrometry, *Anal. Chem.*, *51*:1201 (1979).

5. T. Ito, H. Kawaguchi, and A. Mizuike, Inductively coupled plasma emission spectrometry of microliter samples by flow injection techniques, *Bunseki Kagaku*, *29*:332 (1980).

6. J. T. Vanderslice, G. R. Beecher, and A. G. Rosenfeld, Dispersion and diffusion coefficients in flow injection analysis, *Anal. Chem.*, *56*:292 (1954).

7. J. F. Tyson, Extended calibration of flame atomic-absorption instruments by a flow injection peak width method, *Analyst (London)*, *109*:319 (1984).

8. J. M. H. Appleton and J. F. Tyson, Flow injection atomic absorption spectrometry: the kinetics of instrument response, *J. Anal. At. Spectrom.*, *1*:63 (1986).

9. J. F. Tyson, Flow injection analysis techniques for atomic-absprption spectrometry, *Analyst (London)*, *110*:419 (1985).

10. J. F. Tyson, C. E. Adeeyinwo, J. M. H. Appleton, S. R. Bysouth, A. B. Idris, and L. L. Sarkissian, Flow injection techniques of methods development for flame atomic-absorption spectrometry, *Analyst (London)*, *110*:487 (1985).

11. K. Fukamachi and N. Isibashi, Flow injection. Atomic absorption spectrometry with organic solvents, *Anal. Chim. Acta*, *119*:383 (1980).

12. J. L. Burguera, M. Burguera, M. Gallignani, and O. M. Alarcón, More on flow injection/atomic absorption analysis for electrolytes, *Clin. Chem.*, *29*:568 (1983).

13. A. S. Attiyat and G. D. Christian, Nonaqueous solvents as carrier of sample solvent in flow injection analysis/atomic absorption spectrometry, *Anal. Chem.*, *56*:439 (1984).

14. B. F. Rocks, R. A. Sherwood, and C. Riley, Direct determination of calcium and magnesium in serum using flow-injection analysis and atomic absorption spectroscopy, *Ann. Clin. Biochem.*, *21*:51 (1984).

15. L. Nord and B. Karlberg, Sample preconcentration by continuous flow extraction with a flow injection atomic absorption detection system, *Anal. Chim. Acta*, *145*:51 (1983).

16. J. M. Carter and G. Nickless, Solvent extraction techniques with the technicon autoanalyzer, *Analyst (London)*, *95*:148 (1970).

17. F. Lázaro B., D. Luque de Castro, and M. Valcárcel C., Flow injection environmental analysis: a review, *Analysis*, *13*: 147 (1985).

18. L. Opheim and W. Lund, The use of peristaltic mini-pumps in automated analysis, *Anal. Chim. Acta*, *90*:245 (1977).

19. C. B. Ranger, Flow injection analysis, *Anal. Chem.*, *53*:20A (1981).

20. M. S. Cresser, Discrete sample nebulization in atomic spectroscopy, *Prog. Anal. At. Spectrosc.*, *4*:219 (1981).

21. T. Uchida, I. Kojima, and C. Iida, Determination of metals in small samples by atomic absorption and emission spectrometry with discrete nebulization, *Anal. Chim. Acta*, *116*:205 (1980).

22. B. F. Rocks, R. A. Sherwood, L. M. Bayford, and C. Riley, Zinc and copper determination in microsamples of serum by flow injection and atomic absorption spectrometry, *Ann. Clin. Biochem.*, *19*:338 (1982).

23. J. Růžička and E. H. Hansen, Flow injection analysis. Principles, applications and trends, *Anal. Chim. Acta*, *114*:19 (1982).

24. D. Betteridge, Flow injection analysis, *Anal. Chem.*, *50*:832A (1978).

25. Z. Fang, J. Růžička, and E. H. Hansen, An efficient flow-injection system with on-line ion-exchange preconcentration for the determination of trace amounts of heavy metals by atomic absorption spectrometry, *Anal. Chim. Acta*, *164*:23 (1984).

26. L. Andersson, Simultaneous spectrophotometric determination of nitrite and nitrate by flow injection analysis, *Anal. Chim. Acta*, *110*:123 (1979).

27. C. Riley, B. F. Rocks, R. A. Sherwood, L. H. Aslett, and P. R. Oldfield, A stopped-flow-injection system for automation

of alpha 2 macroglobulin kinetic studies, *J. Autom. Chem.*, 5:32 (1983).

28. C. Riley, B. F. Rocks, and R. A. Sherwood, Controlled-dispersion flow analysis. Flow-injection analysis applied to clinical chemistry, *Anal. Chim. Acta, 179*:69 (1986).

29. F. H. Bergamin F⁰., E. A. G. Zagatto, F. J. Krug, and B. F. Reis, Merging zones in flow injection analysis. Part 1. Double proportional injector and reagent consumption, *Anal. Chim. Acta, 101*:17 (1978).

30. M. Valcárcel and M. D. Lugue de Castro, *Análisis por Invección en Flujo Continuo*, Departamento de Química Analítica, Universidad de Córdoba y Monte Piedad y Caja de Ahorros de Córdoba, Córdoba, Spain, pp. 123-124, 1984.

31. J. Mindegaard, Flow multi-injection analysis. A system for the analysis of highly concentrated samples without prior dilution, *Anal. Chim. Acta, 104*:185 (1979).

32. F. J. Krug, F. H. Bergamin F⁰., and E. A. G. Zagatto, Commutation in flow injection analysis, *Anal. Chim. Acta, 179*:103 (1986).

33. R. A. Sherwood, B. F. Rocks, and C. Riley, Controlled dispersion flow analysis with atomic-absorption detection for the determination of clinically relevant elements, *Analyst (London)*, 110:493 (1985).

34. J. Růžička and E. H. Hansen, Recent developments in flow injection analysis: gradient techniques and hydrodynamic injection, *Anal. Chim. Acta, 145*:1 (1983).

35. P. K. Desgupta and H. Hwang, Application of a nested loop system for the flow injection analysis of trace aqueous peroxides, *Anal. Chem.*, 57:1009 (1985).

36. J. F. Tyson, J. M. H. Appleton, and A. B. Idris, Flow injection sample introduction methods for atomic absorption spectrometry, *Analyst (London)*, 108:153 (1983).

37. S. Greenfield, Inductively coupled plasma-atomic emission spectroscopy (ICP-AES) with flow injection analysis (FIA), *Spectrochim. Acta, Part B, 38*: 93 (1983).

38. J. F. van Staden, A coated tubular solid-stage chloride-selective electrode in flow-injection analysis, *Anal. Chim. Acta, 179*:407 (1986).

39. J. Růžička, Flow injection analysis. A survey of its potential for spectroscopy, *Fresenius Z. Anal. Chem.*, 324:745 (1986).

40. Z. Fang, S. Xu, X. Wang, and S. Zang, Combination of flow-injection techniques with atomic spectrometry for agricultural and environmental analysis, *Anal. Chim. Acta, 179*:325 (1986).

41. G. E. Pacey, M. R. Straka, and J. R. Gord, Dual phase gas diffusion flow injection analysis/hydride generation atomic absorption spectrometry, *Anal. Chem.*, *58*:502 (1986).

42. P. Barrett, L. J. Davidowski, K. W. Penaro, and T. R. Copeland, Microwave oven-based wet digestion technique, *Anal. Chem.*, *50*:1021 (1986).

43. M. Burguera, J. L. Burguera, A. M. Garaboto, and O. M. Alarcón, Determination of iron and copper in infant formula powdered milks by flow injection atomic absorption spectrometry, *Quim. Anal.*, in press (1987).

44. J. L. Burguera, M. Burguera, C. E. Rondón, C. Rivas, J. A. Burguera, and O. M. Alarcón, Determination of lead in hair among gas station exposed workers and in unexposed adults by microwave-aided dissolution of samples and flow injection/atomic absorption spectrometry, *J. Trace Elem. Electrolytes Health Dis.*, *1*:21 (1987).

45. M. Burguera and C. Rondón, Microassay of lead in hair. A microwave-oven decomposition improved digestion procedure followed by flow injection/atomic absorption determination, Proceedings of Heavy Metals in the Environment '87, New Orleans, Vol. 2, p. 35, 1987.

46. M. Burguera and J. L. Burguera, Flow injection and microwave-oven sample decomposition for determination of copper, zinc and iron in whole blood by atomic absorption spectrometry, *Anal. Chim. Acta, 179*:351 (1986).

47. C. W. McLeod, P. J. Worsfold, and A. G. Cox, Simultaneous multi-element analysis of blood serum by flow injection-inductively coupled plasma atomic-emission spectrometry, *Analyst (London), 109*:327 (1984).

48. W. J. Price, *Analytical Atomic Absorption Spectrometry*, Heyden, London, pp. 70-112, 1974.

49. A. Ríos, M. D. Luque de Castro, and M. Valcárcel C., Multidetection in unsegmented flow systems with a single detector, *Anal. Chem., 57*:1803 (1985).

50. W. R. Wolf and J. M. Harnly, Automated simultaneous multi-element flame atomic absorption utilizing flow injection analysis: AMFIA-SIMAAC, Abstracts of the Pittsburgh Conference on Analytical Chemistry and Applied Spectroscopy, Atlantic City, N.J., p. 347, 1980.

51. M. D. Luque de Castro and M. Valcárcel C., Simultaneous determination in flow injection analysis. A review, *Analyst (London), 109*:413 (1984).

52. M. Silva, M. Gallego, and M. Valcárcel C., Sequential atomic absorption spectrometric determination of nitrate and nitrite in meat by liquid-liquid extraction in a flow-injection system, *Anal. Chim. Acta, 179*:341 (1986).

53. S. Olsen, L. C. R. Pessenda, J. Růžička, and E. H. Hansen,
     Combination of flow injection analysis with flame atomic-absorp-
     tion spectrophotometry: determination of trace amounts of
     heavy metals in polluted seawater, Analyst (London), 108:905
     (1983).

54. A. T. Haj-Hussein, G. D. Christian, and J. Růžička, Deter-
     mination of cyanide by atomic absorption using a flow injection
     conversion method, Anal. Chem., 58:38 (1986).

55. J. T. van Gemert, Automated wet chemical analyzer and their
     applications, Talanta, 20:1045 (1973).

56. P. van Der Winkel, G. De Backer, M. Vandeputte, N. Mertens,
     L. Dryon, and D. L. Massart, Performance and characteristics
     of the fluoride-selective electrode in a flow injection system,
     Anal. Chim. Acta, 145:207 (1983).

57. J. Ruíz, A. Ríos, M. D. Luque de Castro, and M. Valcárcel,
     Simultaneous and sequential determination of chromium(VI) and
     chromium(III) by unsegmented flow methods, Fresenius Z. Anal.
     Chem., 322:499 (1985).

58. C. Riley, B. F. Rocks, and R. A. Sherwood, Flow injection
     analysis in clinical chemistry, Talanta, 31:879 (1984).

59. B. F. Rocks, R. A. Sherwood, and C. Riley, Controlled-dis-
     persion flow analysis in clinical chemistry: determination of
     albumin, triglycerides and theophylline, Analyst (London),
     194:847 (1986).

60. N. Zhou, W. Frech, and E. Lundberg, Rapid determination of
     lead, bismuth, antimony and silver in steels by flame atomic
     absorption spectrometry combined with flow injection analysis,
     Anal. Chim. Acta, 153:23 (1983).

61. H. Narasaki and M. Ikeda, Automated determination of arsenic
     and selenium by atomic absorption spectrometry with hydride
     generation, Anal. Chem., 56:2059 (1984).

62. N. H. Tioh, Y. Israel, and R. M. Barnes, Determination of
     arsenic in glycerine by flow injection, hydride generation and
     inductively-coupled plasma/atomic emission spectrometry, Anal.
     Chim. Acta, 184:205 (1986).

63. R. R. Liversage, J. C. van Loon, and J. C. de Andrade, A
     flow injection/hydride generation system for the determination
     of arsenic by inductively coupled plasma atomic emission spec-
     trometry, Anal. Chim. Acta, 161:275 (1984).

64. O. Åström, Flow injection analysis for the determination of
     bismuth by atomic absorption spectrometry with hydride gen-
     eration, Anal. Chem., 54:190 (1982).

65. G. Becerra, J. L. Burguera, and M. Burguera, Determination
     of lead and cadmium in food samples by flow injection atomic
     absorption spectrometry, Quím. Anal., 6: (1987), in press.

66. P. Martínez-Jiménez, M. Gallego, and M. Valcárcel, Indirect determination of cerium and lanthanum by flow injection analysis using an air-acetylene flame, *At. Spectrosc.*, *6*:139 (1985).

67. J. F. Tyson and A. B. Idris, Determination of chromium in steel by flame atomic-absorption spectrometry using a flow injection standard additions method, *Analyst (London)*, *109*:23 (1984).

68. E. A. G. Zagatto, A. O. Jacintho, F. J. Krug, B. T. Reis, A. E. Bruns, and M. C. U. Araújo, Flow injection systems with inductively-coupled argon plasma atomic emission spectrometry. Part 2. The generalized standard addition method, *Anal. Chim. Acta*, *145*:169 (1983).

69. C. G. Taylor and J. M. Trevaskis, Determination of lead in gasoline by a flow-injection technique with atomic absorption spectrometric detection, *Anal. Chim. Acta*, *179*:491 (1986).

70. M. Gallego, M. Silva, and M. Valcárcel, Determination of nitrate and nitrite by continuous liquid-liquid extraction with a flow-injection atomic-absorption detection system, *Fresenius Z. Anal. Chem.*, *323*:50 (1986).

71. Y. Israel and R. M. Barnes, Standard addition method in flow injection analysis with inductively coupled plasma atomic emission spectrometry, *Anal. Chem.*, *56*:1192 (1984).

72. P. Martínez-Jiménez, M. Gallego, and M. Valcárcel, Indirect atomic absorption determination of uranium by flow injection analysis using an air-acetylene flame, *At. Spectrosc.*, *6*:65 (1985).

Appendix A: List of Symbols

A	absorbance
C	concentration (mol L^{-1})
C_A	mass transport of a component A
C_i, C_j	concentration of the ith and jth samples (mol L^{-1})
C_s	concentration of the standard solution (mol L^{-1})
C_t	concentration input function
C_x	concentration of sample solution (mol L^{-1})
D	dispersion (C^o/C_{max}) or diffusion coefficient ($m^2\ s^{-1}$)
$\tilde{D}$	nonisotropic dispersion coefficient
De	Dean number
D_L	axial dispersion coefficient
D_R	radial dispersion coefficient
F	mean linear flow velocity ($Q\pi^{-1}r^{-2}$ cm s^{-1})
f_v	volumetric flow rate
H	height of analytical signal (mm)
hd(τ)	impulse response function
I	peak height of the signal (mm)
IL	injector length
L	length of the tube (m)
Pe	Péclet number

Pe_L longitudinal Péclet number

Pe_R radial Péclet number

R   conduit radius (m)

R_A source or sink term accounting for the production or disappearance of a component

R_{coil} radius of the coil

r   internal radius

Re   Reynolds number

Sc   Schmidt number

S_{max} maximum sample frequency

Sv   sample volume

t   time(s)

t_A time elapsed between injection and the initial appearance of a peak at the detector

t_B peak width (baseline to baseline)

t_v residence time

$\bar{v}$ velocity vector

$<v>$ mean linear flow velocity

x   characteristic length

θ dynamic viscosity (P_a s)

η concentration obtained when all the injected components are distributed homogeneously along a tube

ρ density (kg/m^3)

ν kinematic viscosity ($m^2 \, s^{-1}$)

Appendix B: FIA-AS Bibliography

1. P. W. Alexander, R. J. Finlayson, L. E. Smythe, and A. Thalib, Rapid flow analysis with inductively coupled plasma atomic-emission spectroscopy using a micro-injection technique, *Analyst, 107*: 1335 (1982).

2. J. Alonso, J. Bartroli, J. L. F. C. Lima, and A. A. S. C. Machado, Sequential flow injection determination of calcium and magnesium in waters, *Anal. Chim. Acta, 179*: 503 (1986).

3. J. M. H. Appleton and J. F. Tyson, Flow injection atomic absorption spectrometry: the kinetics of instrument response, *J. Anal. At. Spectrom., 1*: 63 (1986).

4. M. C. U. Araujo, C. Pasquini, R. E. Bruns, and E. A. G. Zagatto, A fast procedure for standard additions in flow injection analysis, *Anal. Chim. Acta, 171*: 337 (1985).

5. O. Astrőm, Flow injection analysis for the determination of bismuth by atomic absorption spectrometry with hydride generation, *Anal. Chem., 54*: 190 (1982).

6. A. Attiyat and G. D. Christian, Nonaqueous solvents as carrier or sample solvent in flow injection analysis/atomic absorption spectrometry, *Anal. Chem., 56*: 439 (1984).

7. A. Attiyat, Study of sample solvent/carrier combination for flow injection analysis-atomic absorption spectrometry, *J. of Flow Injection Analysis, 4*: 26 (1987).

8. A. S. Attiyat and G. D. Christian, Flow injection analysis—atomic absorption determination of serum zinc, *Clinica. Chim. Acta, 137*: 151 (1984).

9. K. Backstrőm, L. G. Danielsson, and L. Nord, Sample work-up for graphite furnace atomic-absorption spectrometry using continuous flow extraction, *Analyst, 109*: 323 (1984).

The editor wishes to thank Professor Kent K. Stewart, author of Chapter 1, for having compiled this comprehensive listing of references on FIA-AS.

10. W. D. Basson and J. F. van Staden, Simultaneous determination of sodium, potassium, magnesium and calcium in surface, ground, and domestic water by flow-injection analysis, *Fres. Z. Anal. Chem.*, *302*: 370 (1980).

11. M. Bengtsson and G. Johansson, Preconcentration and matrix isolation of heavy metals through a two stage solvent and extraction in a flow system, *Anal. Chim. Acta*, *158*: 147 (1984).

12. M. Bengtsson, F. Malamas, A. Torstensson, O. Regnell, and G. Johansson, Trace metal ion preconcentration for flame atomic absorption by an immobilized N,N,N'-tri-(2-pyridylmethyl) ethylenediamine (Tri-PEN) chelate ion exchanger in a flow injection system, *Mikrochim. Acta*, *3*: 209 (1985).

13. J. A. C. Broekaert and F. Leis, An injection method for the sequential determination of boron and several metals in wastewater samples by inductively-coupled plasma atomic emission spectrometry, *Anal. Chim. Acta*, *109*: 73 (1979).

14. M. W. Brown and J. Růžička, Parameters affecting sensitivity and precision in the combination of flow injection analysis with flame atomic-absorption spectrophotometry, *Analyst*, *109*: 1091 (1984).

15. R. F. Browner, Sample introduction for inductively coupled plasmas and flames, *Trends in Anal. Chem.*, *2*: 121 (1983).

16. R. F. Browner and A. W. Boorn, Sample introduction: the Achilles' heel of atomic spectroscopy, *Anal. Chem.*, *56*: 786A (1984).

17. R. F. Browner and A. W. Boorn, Sample introduction: techniques for atomic spectroscopy, *Anal. Chem.*, *56*: 875A (1984).

18. M. Burguera, J. Burguera, P. C. Rivas, and O. M. Alarcón, Determination of copper, zinc, and iron in parotid saliva by flow injection with flame atomic absorption spectrophotometry, *At. Spectrosc.*, *7*: 79 (1986).

19. J. L. Burguera and M. Burguera, Flow injection spectrophotometry followed by atomic absorption spectrometry for the determination of iron(II) and total iron, *Anal. Chim. Acta*, *161*: 375 (1984).

20. J. L. Burguera, M. Burguera, and O. M. Alarcón, Determination of sodium, potassium, calcium, magnesium, iron, copper, and zinc in cerebrospiral fluid by flow injection atomic absorption spectrometry, *J. Anal. At. Spectrom.*, *1*: 79 (1986).

21. J. L. Burguera, M. Burguera, M. Gallignani, and O. M. Alarcón, More on flow injection atomic absorption analysis for electrolytes, *Clin. Chem.*, *29*: 568 (1983).

22. J. L. Burguera, M. Burguera, and M. Gallignani, Direct determination of sodium and potassium in blood serum by flow injection and atomic absorption spectrophotometry, *An. Acad. Bras. Cienc.*, *55*: 209 (1983).

23. J. L. Burguera, M. Burguera, L. La Cruz O, and R. Naranjo, Determination of lead in the urine of exposed and unexposed adults by extraction and flow-injection/atomic absorption spectrometry, *Anal. Chim. Acta, 186*: 273 (1986).

24. M. Burguera and J. L. Burguera, Flow injection and microwave-oven sample decomposition for determination of copper, zinc and iron in whole blood by atomic absorption spectrometry, *Anal. Chim. Acta, 179*: 351 (1986).

25. N. D. Bylington, Flow injection atomic absorption assay of copper and zinc in the plasma of age dependent audiogenic seizure susceptible mice, *Diss. Abst. Int., 43/10*: 3228 B (1983).

26. S. R. Bysouth and J. F. Tyson, A microcomputer-based peak-width method of extended calibration for flow-injection atomic absorption spectrometry, *Anal. Chim. Acta, 179*: 481 (1986).

27. C. C. Y. Chan, Semiautomated method for determination of selenium in geological materials using a flow injection analysis technique, *Anal. Chem., 57*: 1482 (1985).

28. J. T. Dobbins and J. M. Martin, Flow injection analysis: twelve years old and growing, *Spectrosc., 1/9*: 20 (1986).

29. F. S. Chuang, J. R. Sarbeck, P. A. St. John, and J. D. Winefordner, Flame spectrometric determination of sodium, potassium and calcium in blood serum by measurement of micro-samples, *Mikrochim. Acta [Wein] 1973*: 523 (1973).

30. I. G. Cook, C. W. McLeod, and P. J. Worsfold, Use of activated alumina as a column packing material for the absorption of oxyanions in flow injection analysis with ICP-AES detection, *Anal. Proc., 23*: 5 (1986).

31. A. G. Cox, I. G. Cook, and C. W. McLeod, Rapid sequential determination of chromium(III)-chromium(VI) by flow injection analysis-inductively coupled plasma atomic-emission spectrometry, *Analyst, 110*: 331 (1985).

32. A. G. Cox and C. W. McLeod, Preconcentration and determination of trace chromium(III) by flow injection/inductively-coupled plasma/atomic emission spectrometry, *Anal. Chim. Acta, 179*: 487 (1986).

33. M. S. Cresser, Discrete sample nebulization in atomic spectroscopy, *Prog. Anal. At. Spectrosc., 4*: 219 (1981).

34. L. G. Dannielson and L. Nord, Sample workup for atomic absorption spectrometry using flow injection extraction, *Anal. Proc., 20*: 298 (1983).

35. D. E. Davey, The analysis of galvanizing preflux solutions for zinc by flow injection-flame atomic absorption spectrometry, *Anal. Lett., 19*: 1573 (1986).

36. J. C. de Andrade, C. Pasquini, N. Baccan, and J. C. van Loon, Cold vapor atomic absorption determination of mercury by flow injection analysis using a Teflon membrane phase separator coupled to the absorption cell, *Spectrochim. Acta,* *38B*: 1329 (1983).

37. Z. Fang, S. Xu, X. Wang, and S. Zhang, Combined flow injection analysis and atomic absorption spectroscopy, *Guangpuxue Yu Gangpu Fenxi, 6*: 31 (1986).

38. Z. Fang, Flow injection techniques in atomic spectroscopy, *Fenxi Huaxue, 14*: 549 (1986).

39. Z. Fang, J. M. Harris, J. Růžička, and E. H. Hansen, Simultaneous flame photometric determination of lithium, sodium, potassium, and calcium by flow injection analysis with gradient scanning standard addition, *Anal. Chem., 57*: 1457 (1985).

40. Z. Fang, S. Xu, and S. Zhang, Determination of trace amounts of nickel by on-line flow injection ion-exchange preconcentration atomic absorption spectrometry, *Fenxi Huaxue, 12*: 997 (1984).

41. Z. Fang, J. Růžička, and E. H. Hansen, An efficient flow-injection system with on-line ion-exchange preconcentration for the determination of trace amounts of heavy metals by atomic absorption spectrometry, *Anal. Chim. Acta, 164*: 23 (1984).

42. Z. Fang, S. Ku, X. Wang, and S. Zhang, Combination of flow-injection techniques with atomic spectrometry in agricultural and environmental analysis, *Anal. Chim. Acta, 179*: 325 (1986).

43. Z. Fang, S. Xu, and S. Zhang, The determination of trace amounts of heavy metals in water by a flow-injection system including ion-exchange preconcentration and flame atomic absorption spectrometric detection, *Anal. Chim. Acta, 164*: 41 (1984).

44. K. Fukamachi and N. Ishibashi, Flow injection-atomic absorption spectrometry with organic solvents, *Anal. Chem. Acta, 119*: 383 (1980).

45. M. Gallego, M. Silva, and M. Valcárcel, Determination of nitrate and nitrite by continuous liquid-liquid extraction with a flow-injection atomic-absorption detection system, *Fres. Z. Anal. Chem., 323*: 50 (1986).

46. M. Gallego and M. Valcárcel, Indirect atomic absorption spectrometric determination of perchlorate by liquid-liquid extraction in a flow-injection system, *Anal. Chim. Acta, 169*: 161 (1985).

47. M. Gallego, M. Silva, and M. Valcárcel, Indirect atomic absorption determination of anionic surfactants in waste-waters by flow injection continuous liquid-liquid extraction, *Anal. Chem., 58*: 2265 (1986).

48. M. Gallignani, J. L. Burguera, and M. Burguera, Determination of calcium and magnesium in blood sera by flow injection analysis and atomic absorption spectrometry, *Acta Cient. Venez.*, *33*: 371 (1982).

49. I. L. Garcia, M. H. Córdoba, and C. Sanchez-Pedreno, Flow injection atomic absorption spectrometry with air compensation, *Analyst*, *112*: 271 (1987).

50. S. Greenfield, Inductively coupled plasma-atomic emission spectroscopy (ICP-AES) with flow injection analysis (FIA), *Spectrochim. Acta, 38B*: 93 (1983).

51. A. M. Gunn, An automated hydride generation atomic absorption spectrometric method for the determination of total arsenic in raw and portable waters, Tech. Rept. TR191, Water Res. Ctr., p. 31 (1983).

52. A. T. Haj-Hussein, G. D. Christian, and J. Růžička, Determination of cyanide by atomic absorption using a flow injection conversion method, *Anal. Chem.*, *58*: 38 (1986).

53. J. M. Harnly and G. R. Beecher, Two-valve injector to minimize nebulizer memory for flow injection atomic absorption spectrometry, *Anal. Chem.*, *57*: 2015 (1985).

54. J. M. Harnly and G. R. Beecher, Signal to noise ratios for flow injection atomic absorption spectrometry, *J. of Anal. At. Spectrom.*, *1*: 75 (1986).

55. S. D. Hartenstein, G. D. Christian, and J. Růžička, Applications of an on-line preconcentration flow injection analysis system for inductively coupled plasma atomic emission spectrometry, *Can. J. Spectrosc.*, *30*: 144 (1985).

56. S. D. Hartenstein, J. Růžička, and G. D. Christian, Sensitivity enhancement for flow injection analysis-inductively coupled plasma atomic emission spectrometry using an on-line preconcentrating ion-exchange column, *Anal. Chem.*, *57*: 21 (1985).

57. P. Herandez, L. Herandez, J. Vicente, and M. T. Sevilla, Use of an FIA ion exchanger-atomic absorption system for determining manganese, lead and copper, *An. Quim., Ser. B, 81*: 117 (1985).

58. P. Herandez, L. Herandez, and J. Losada, Determination of aluminium in hemodialysis fluids by a flow injection system with preconcentration on a synthetic chelate-forming resin and flame atomic absorption spectrophotometry, *Fres. Z. Anal. Chem.*, *325*: 300 (1986).

59. S. Hirata and K. Honda, Rapid determination of zinc in seawater by column preconcentration/AAS, *Bunseki Kagaku, 36*: 213 (1987).

60. S. Hirata, Y. Umezaki, and M. Ikeda, Determination of cadmium of ppb level by column preconcentration-atomic absorption spectrometry, *J. Flow Injection Analysis, 3(1)*: 8 (1986).

61. S. Hirata, Y. Umezaki, and M. Ikeda, Determination of chromium(III), titanium, vanadium, iron(III), and aluminum by inductively coupled plasma atomic emission spectrometry with an on-line preconcentrating ion-exchange column, *Anal. Chem.*, *58*: 2602 (1986).

62. M. Ikeda, Determination of selenium by atomic absorption spectrometry with miniaturized suction-flow hydride generation and on-line removal of interferences, *Anal. Chim. Acta*, *170*: 217 (1985).

63. M. Ikeda, Determination of arsenic at the picogram level by atomic absorption spectrometry with miniaturized suction-flow hydride generation, *Anal. Chim. Acta*, *167*: 289 (1985).

64. Y. Israel and R. M. Barnes, Standard addition method in flow injection analysis with inductively coupled plasma emission spectrometry, *Anal. Chem.*, *56*: 1188 (1984).

65. T. Ito, H. Kawaguchi, and A. Mizuike, Inductively coupled plasma emission spectrometry of microliter samples by a flow injection technique, *Bunseki Kagaku*, *29*: 332 (1980).

66. T. Ito, E. Nakagawa, H. Kawaguchi, and A. Mizuike, Semiautomatic microliter sample injection into an inductively coupled plasma for simultaneous multielement analysis, *Mikrochim. Acta*, *1*: 423 (1982).

67. T. Iwachido, M. Onoda, and S. Motomizu, Determination of potassium in river water by solvent extraction-flow injection analysis, *Anal. Sci.*, *2*: 493 (1986).

68. A. O. Jacintho, E. A. G. Zagatto, H. Bergamin, F⁰., F. J. Krug, and B. F. Reis, Flow injection system with inductively-coupled argon plasma atomic emission spectrometry, Part 1, Fundamental considerations, *Anal. Chim. Acta*, *130*: 243 (1981).

69. O. F. Kamson and A. Townshend, Ion-exchange removal of some interferences on the determination of calcium by flow injection analysis and atomic absorption spectrometry, *Anal. Chim. Acta*, *155*: 253 (1983).

70. E. D. Katz and R. P. W. Scott, Peak dispersion in a liquid chromatography-atomic-absorption spectrometry system, *Analyst*, *110*: 253 (1985).

71. H. Kimura, K. Oguma, and R. Kuroda, Atomic absorption spectrophotometry determination of calcium silicate rocks by a flow injection method, *Bunseki Kagaku*, *32*: T79 (1983).

72. J. A. Koropchak and D. H. Winn, Thermospray interfacing for flow injection analysis with inductively coupled plasma atomic emission spectrometry, *Anal. Chem.*, *58*: 2558 (1986).

73. T. Kumamaru, H. Matsuo, Y. Okamoto, and M. Ikeda, Sensitivity enhancement for inductively-coupled plasma atomic emission spectrometry of cadmium by suction-flow on-line ion-exchange preconcentration, *Anal. Chim. Acta*, *181*: 271 (1986).

74. T. Kumamaru, Y. Nitta, F. Nakata, H. Matsuo, and M. Ikeda, Determination of cadmium by suction-flow liquid-liquid extraction combined with inductively-coupled plasma atomic emission spectrometry, *Anal. Chim. Acta, 174*: 183 (1985).

75. K. E. LaFreniere, G. W. Gray, and V. A. Fassel, Flow injection analysis with inductively coupled plasma-atomic emission spectroscopy: critical comparison of conventional pneumatic, ultrasonic and direct injection nebulization, *Spectrochim. Acta, Part B, 40*: 1495 (1985).

76. K. E. Lawrence, G. W. Rice, and V. A. Fassel, Direct liquid sample introduction for flow injection analysis and liquid chromatography with inductively coupled argon plasma spectrometric detection, *Anal. Chem., 56*: 289 (1984).

77. N. Leon, J. L. Burguera, M. Burguera, and O. M. Alarcón, Determination of cobalt and manganese in blood serum by flow injection analysis and atomic absorption spectroscopy, *Rev. Roum. Chim., 31*: 353 (1986).

78: R. R. Liversage, J. C. van Loon, and J. C. de Andrade, A flow injection/hydride generation system for the determination of arsenic by inductively-coupled plasma atomic emission spectrometry, *Anal. Chim. Acta, 161*: 275 (1984).

79. M. D. Luque de Castro, Flow injection analysis: a new tool to automate extraction processes, *J. Auto. Chem., 8*: 56 (1986).

80. M. D. Luque de Castro and M. Valcárcel, Simultaneous determinations in flow analysis, A review, *Analyst, 109*: 413 (1984).

81. T. P. Lynch, N. J. Kernoghan, and J. N. Wilson, Speciation of metals in solution by flow injection analysis, Part 1, Sequential spectrophotometric and atomic-absorption detectors, *Analyst, 109*: 839 (1984).

82. F. Malamas, M. Bengtsson, and G. Johansson, On-line trace metal enrichment and matrix isolation in atomic absorption spectrometry by a column containing immobilized 8-quinolinol in a flow-injection system, *Anal. Chim. Acta, 160*: 1 (1984).

83. M. A. Marshall and H. A. Mottola, Performance studies under flow conditions of silica-immobilized 8-quinolinol and its application as a preconcentration tool in flow injection/atomic absorption determinations, *Anal. Chem., 57*: 729 (1985).

84. P. Martinez-Jimenez, M. Gallego, and M. Valcárcel, Indirect atomic absorption determination of uranium by flow injection analysis using an air-acetylene flame, *At. Spectroc., 6*: 65 (1983).

85. P. Martinez-Jimenez, M. Gallego, and M. Valcárcel, Indirect atomic absorption determination of cerium and lanthanum by flow injection analysis using an air-acetylene flame, *At. Spectroc. 6*: 137 (1985).

86. P. Martinez-Jimenez, M. Gallego, and M. Valcárcel, Indirect atomic absorption determination of aluminum by flow injection analysis, *Microchem. J.*, *34*: 190 (1986).

87. P. Martinez-Jimenez, M. Gallego, and M. Valcárcel, Analytical potential of continuous precipitation in flow injection-atomic absorption configurations, *Anal. Chem.*, *59*: 69 (1987).

88. C. W. McLeod, P. J. Worsfold, and A. G. Cox, Simultaneous multi-element analysis of blood serum by flow injection-inductively coupled plasma atomic-emission spectrometry, *Analyst*, *109*: 327 (1984).

89. C. W. McLeod, I. G. Cook, P. J. Worsfold, J. E. Davies, and J. Queay, Analyte enrichment and matrix removal in flow injection analysis-inductively coupled plasma-atomic emission spectrometry: determination of phosphorus in steels, *Spectrochim. Acta*, *40B*: 57 (1985).

90. E. B. Milosavljevic, J. Růžička, and E. H. Hansen, Simultaneous determination of free and EDTA-complexed copper ions by flame atomic absorption with an ion-exchange flow-injection system, *Anal. Chim. Acta*, *169*: 321 (1985).

91. B. D. Mindel and B. Karlberg, A sample pretreatment system for atomic absorption using flow injection analysis, *Lab. Proc.*, *129*: 720 (1981).

92. H. Morita, T. Kimoto, and S. Shimomura, Flow injection analysis of mercury/cold vapor atomic fluorescence spectrophotometry, *Anal. Lett.*, *16(A15)*: 1187 (1983).

93. H. Narasaki and M. Ikeda, Automated determination of arsenic and selenium by atomic absorption spectrometry with hydride generation, *Anal. Chem.*, *56*: 2059 (1984).

94. L. Nord and B. Karlberg, An automated extraction system for flame atomic absorption spectrometry, *Anal. Chim. Acta*, *125*: 199 (1981).

95. L. Nord and B. Karlberg, Sample preconcentration by continuous flow extraction with a flow injection atomic absorption detection system, *Anal. Chim. Acta*, *145*: 151 (1983).

96. K. Ogata, S. Tamabe, and T. Imanari, Flame atomic absorption spectrophotometry coupled with solvent extraction/flow injection analysis, *Chem. Pharm. Bull.*, *31*: 1419 (1983).

97. K. Oguma, T. Nara, and R. Kuroda, Atomic absorption spectrophotometric determination of magnesium in silicates by flow injection method, *Bunseki Kagaku*, *35*: 690 (1986).

98. J. W. Olesik and S. V. Olesik, Supercritical fluid-based sample introduction for inductively coupled plasma atomic spectrometry, *Anal. Chem.*, *59*: 796 (1987).

99. S. Olsen, Trace analysis of heavy metals in seawater, flow injection analysis and atomic absorption spectrophotometry, *Dan. Kemi*, *64*: 68 (1983).

100. S. Olsen, L. C. R. Pessenda, J. Růžička, and E. H. Hansen, Combination of flow injection analysis with flame atomic-absorption spectrophotometry: determination of trace amounts of heavy metals in polluted seawater, *The Analyst, 108*: 905 (1983).

101. G. E. Pacey, M. R. Straka, and J. R. Gord, Dual phase gas diffusion flow injection analysis/hydride generation atomic absorption spectrometry, *Anal. Chem., 58*: 502 (1986).

102. G. E. Pacey and B. P. Bubnis, Flow injection analysis as a tool for metal speciation, *Amer. Lab.*, July 17, 1984.

103. K. Y. Patterson, G. R. Beecher, W. R. Wolf, and J. C. J. R. Smith, Analysis of small volume samples by flow injection analysis flame atomic absorption (FIAAA), *Abst. Pitt. Con.*: 421 (1980).

104. B. A. Petersson, Z. Fang, J. Růžička, and E. H. Hansen, Conversion techniques in flow injection analysis determination of sulphide by precipitation with cadmium ions and detection by atomic absorption spectrometry, *Anal. Chim. Acta, 184*: 165 (1986).

105. J. Pettersson, L. Hansson, and A. Olin, Comparison of four digestion methods for the determination of selenium in bovine liver by hydride generation and atomic-absorption spectrometry in a flow system, *Talanta, 33*: 249 (1986).

106. R. Playle, J. Gleed, R. Jonasson, and J. R. Kramer, Comparison of atomic absorption spectrometric, spectrophotometric, and fluorimetric methods for determination of aluminum in water, *Anal. Chim. Acta, 134*: 369 (1982).

107. M. J. Powell, D. W. Boomer, and R. J. McVicars, Introduction of gaseous hydrides into an inductively coupled plasma mass spectrometer, *Anal. Chem., 58*: 2864 (1986).

108. M. H. Ramsey and M. Thompson, One-line dilutor for atomic absorption spectrometry, *Analyst, 107*: 232 (1982).

109. B. F. Reis, A. O. Jacintho, J. Mortatti, F. J. Krug, and E. A. G. Zagatto, Zone-sampling processes in flow injection analysis, *Anal. Chim. Acta, 123*: 221 (1981).

110. B. W. Renoe and A. O'Brien, Calcium by flow injection atomic absorption, *Clin. Chem., 26*: 1021 (1980).

111. B. W. Renoe, C. E. Shildeler, and J. Savory, Use of a flow-injection sample manipulator as an interface between a "high performance" liquid chromatograph and an atomic absorption spectrophotometer, *Clin. Chem., 27*: 1546 (1981).

112. C. Riley, B. F. Rocks, and R. A. Sherwood, Flow injection analysis in clinical chemistry, *Talanta, 31*: 879 (1984).

113. B. F. Rocks, R. A. Sherwood, L. M. Bayford, and C. Z.
      Riley, Zinc and copper determination in microsamples of serum
      by flow injection and atomic absorption spectroscopy, *Ann.
      Clin. Biochem.*, *19*: 338 (1982).

114. B. F. Rocks, R. A. Sherwood, and C. Riley, Direct determin-
      ation of therapeutic concentrations of lithium in serum by flow-
      injection analysis with atomic absorption spectroscopic detection,
      Clin. Chem., *28*: 440 (1982).

115. B. F. Rocks, R. A. Sherwood, and C. Riley, *Cim. Chem.*,
      29: 569 (1983), letter.

116. B. F. Rocks, R. A. Sherwood, and C. Riley, Direct determin-
      ation of calcium and magnesium in serum using flow-injection
      analysis and atomic absorption spectroscopy, *Ann. Clin. Bio-
      chem.*, *21*: 51 (1984).

117. B. F. Rocks, R. A. Sherwood, Z. J. Turner, and C. Riley,
      Serum iron and total iron-binding capacity determination by
      flow-injection analysis with atomic absorption detection, *Ann.
      Clin. Biochem.*, *20*: 72 (1983).

118. J. Růžička, Flow injection analysis, A survey of its potential
      for spectroscopy, *Fres. Z. Anal. Chem.*, *324*: 745 (1986).

119. J. Růžička and E. H. Hansen, *Flow Injection Analysis, Chem-
      ical Analysis*, Vol. 62, Wiley, New York, 1981.

120. J. Růžička and E. H. Hansen, Recent development in flow in-
      jection analysis: gradient techniques and hydrodynamic injec-
      tion, *Anal. Chim. Acta*, *145*: 1 (1983).

121. J. R. Sarbeck, P. A. St. John, and J. D. Winefordner, Mea-
      surement of microsamples in atomic emission and atomic fluores-
      cence flame spectrometry, *Mikrochim. Acta [Wien]*: 55 (1972).

122. R. A. Sherwood, B. F. Rocks, and C. Riley, Controlled-dis-
      persion flow analysis with atomic-absorption detection for the
      determination of clinically relevant elements, *Analyst*, *110*:
      493 (1985).

123. F. Shirato, Y. Okajima, T. Kuroishi, and Y. Takata, Deter-
      mination of Ni in Ni-Fe alloy thin films by FIA, *Bunseki Kagaku*,
      36: 233 (1987).

124. L. K. Shpigun, I. Y. Kolotyrkina, and Y. A. Zolotov, Flow-
      injection analysis, spectrophotometric determination of nickel,
      Zh. Anal. Chim., *41*: 1224 (1986).

125. M. Silva, M. Gallego, and M. Valcárcel, Sequential atomic ab-
      sorption spectrometric determination of nitrate and nitrite in
      meats by liquid-liquid extraction in a flow-injection system,
      Anal. Chim. Acta, *179*: 341 (1986).

126. K. K. Stewart and A. G. Rosenfeld, Exponential dilution cham-
      bers for scale expansion in flow injection analysis, *Anal. Chem.*,
      54: 2368 (1982).

127. J. A. Sweileh and F. F. Cantwell, Sample introduction by solvent extraction/flow injection to eliminate interferences in atomic absorption spectroscopy, *Anal. Chem.*, *57*: 420 (1985).

128. C. G. Taylor and J. Trevaskis, Determination of lead in gasoline by a flow-injection technique with atomic absorption spectrometric detection, *Anal. Chim. Acta*, *179*: 491 (1986).

129. J. J. Thompson and R. S. Houk, Inductively coupled plasma mass spectrometric detection for multielement flow injection analysis and elemental speciation by reversed-phase liquid chromatography, *Anal. Chem.*, *58*: 2541 (1986).

130. N. H. Tioh, Y. Israel, and R. M. Barnes, Determination of arsenic in glycerine by flow injection analysis, hydride generation and inductively-coupled plasma/atomic emission spectrometry, *Anal. Chim. Acta*, *184*: 205 (1986).

131. A. Townshend, Ion-exchange minicolumns, *Anal. Chim. Acta*, *180*: 49-51 (1986).

132. J. Tyson, Flow injection techniques for flame atomic absorption spectrophotometry, *Trends in Anal. Chem.*, *4*: 124 (1985).

133. J. F. Tyson, J. M. H. Appleton, and A. B. Idris, Flow injection calibration methods for atomic absorption spectrometry, *Anal. Chim. Acta*, *145*: 159 (1983).

134. J. F. Tyson and A. B. Idris, Flow injection sample introduction for atomic-absorption spectrometry: applications of a simplified model for dispersion, *Analyst*, *106*: 1125 (1981).

135. J. Tyson, Flow injection analysis combined with atomic-absorption spectrometry, *Anal. Proc.*, *21*: 377 (1984).

136. J. F. Tyson and J. M. H. Appleton, Flow injection calibration methods for flame atomic absorption spectrometry, *Anal. Proc.*, *20*: 260 (1983).

137. J. F. Tyson, Flow injection methods and atomic-absorption spectrophotometry, *Anal. Proc.*, *20*: 488 (1983).

138. J. F. Tyson, Extended calibration of flame atomic-absorption instruments by a flow injection peak width method, *Analyst*, *109*: 319 (1984).

139. J. F. Tyson, Flow injection analysis techniques for atomic-absorption spectrometry: a review, *Analyst*, *110*: 419 (1985).

140. J. F. Tyson, Flow injection techniques for extending the working range of atomic absorption spectrometry and U.V. visible spectrometry, *Anal. Chim. Acta*, *180*: 51 (1986).

141. J. F. Tyson, C. E. Adeeyinwo, J. M. H. Appleton, S. R. Bysouth, A. B. Idris, and L. L. Sarkissian, Flow injection techniques of method development for flame atomic-absorption spectrometry, *Analyst*, *110*: 487 (1985).

142. J. F. Tyson, J. M. H. Appleton, and A. B. Idris, Flow in-
     jection samples—introduction methods for atomic-absorption
     spectrometry, Analyst, 108: 153 (1983).

143. J. F. Tyson and A. B. Idris, Determination of chromium in
     steel by flame atomic-absorption spectrometry using a flow in-
     jection standard additions method, Analyst, 109: 23 (1984).

144. T. Uchida, C. Wei, C. Iida, and H. Wada, Simultaneous de-
     termination of calcium and magnesium in serum with flow injec-
     tion-atomic absorption system, Nagoya Kogyo, 33: 97 (1982).

145. M. Valcárcel, M. Gallego, and P. Martinez-Jimenez, Indirect
     atomic absorption methods based on continuous precipitation
     in flow injection analysis, Anal. Proc., 23: 233 (1986).

146. M. Valcárcel and M. D. Luque de Castro, Flow Injection Ana-
     lysis—Principles and Applications, Wiley, New York, 1987
     (English ed.).

147. X. Wang and Z. Fang, Determination of trace amounts of se-
     lenium in environmental samples by hydride generation-atomic
     absorption spectrometry combined with flow injection analysis
     techniques, Kexue Tongboa (foreign lang. ed.), 31: 791
     (1986).

148. W. R. Wolf and K. K. Stewart, Automated multiple flow injec-
     tion analysis for flame atomic absorption spectrometry, Anal.
     Chem., 51: 1201 (1979).

149. W. R. Wolf and J. M. Harnly, Automated simultaneous multi-
     element flame atomic absorption utilizing flow injection analysis:
     AMFIA SIMAAC, Abstr. Pitts. Con.: 347 (1980).

150. M. Yamamoto, M. Yasuda, and Y. Yamamoto, Flow injection-
     hydride generation-atomic absorption spectrometry with gas
     diffusion unit using microporous PTFE tube, J. Flow Injection
     Analysis, 2: 134 (1985).

151. M. Yamamoto, M. Yasuda, and Y. Yamamoto, Hydride-genera-
     tion atomic absorption spectrometry coupled with flow injection
     analysis, Anal. Chem., 57: 1382 (1985).

152. N. Yoza, Y. Aoyagi, and S. Ohashi, Flow injection system for
     atomic absorption spectrometry, Anal. Chim. Acta, 111: 163 (1979).

153. E. A. G. Zagatto, O. Bania, and M. F. Gine, A simple proce-
     dure for hydrodynamic injection in flow injection analysis ap-
     plied to the atomic absorption spectrometry of chromium in
     steels, Anal. Chim. Acta, 181: 265 (1986).

154. E. A. G. Zagatto, M. F. Gine, E. A. N. Fernandes, B. F.
     Reis, and F. J. Krug, Sequential injections in flow systems
     as an alternative to gradient exploitation, Anal. Chim. Acta,
     173: 289 (1985).

155. E. A. G. Zagatto, A. O. Jacintho, F. J. Krug, and B. F. Reis, Flow injection systems with inductively-coupled argon plasma atomic emission spectrometry, Part 2, The generalization standard addition method, *Anal. Chim. Acta, 145*: 169 (1983).

156. E. A. G. Zagatto, F. J. Krug, H. Bergamin, F⁰., S. S. Jørgensen, and B. F. Reis, Merging zones in flow injection analysis, Part 2, Determination of calcium, magnesium and potassium in plant material by continuous flow injection atomic absorption and flame emission spectrometry, *Anal. Chim. Acta, 104*: 279 (1979).

157. S. Zhang, L. Sun, H. Jiang, and Z. Fang, Determination of copper, zinc, iron, manganese, sodium, potassium, calcium and magnesium in plants and soil by flow injection atomic absorption spectrometry, *Guangpuxue Yu Guangpu Fenxi, 4*: 42 (1984).

158. N. Zhou, W. Frech, and E. Lundberg, Rapid determination of lead, bismuth, antimony and silver in steels by flame atomic absorption spectrometry combined with flow injection analysis, *Anal. Chim. Acta, 153*: 23 (1983).

Author Index

Subject Index